NB// Loan label on next page

Sheep Management and Production

A practical guide for farmers and students

Derek H. Goodwin
N.D.A., S.C.D.A., F.T.C. (Agric.), Cert. Ed. (Birmingham)

Lecturer in Animal Husbandry
Gloucestershire College of Agriculture

Hutchinson
London Melbourne Sydney Auckland Johannesburg

Hutchinson & Co. (Publishers) Ltd

An imprint of the Hutchinson Publishing Group

17–21 Conway Street, London W1P 6JD

Hutchinson Group (Australia) Pty Ltd
30–32 Cremorne Street, Richmond South, Victoria 3121
PO Box 151, Broadway, New South Wales 2007

Hutchinson Group (NZ) Ltd
32–34 View Road, PO Box 40 086, Glenfield, Auckland 10

Hutchinson Group (SA) (Pty) Ltd
PO Box 337, Bergvlei 2012, South Africa

First published as *The Production and Management of Sheep* 1971
Reprinted 1974, 1977
Second edition 1979
Reprinted 1980, 1981, 1982

Printed in Great Britain by The Anchor Press Ltd
and bound by Wm Brendon & Son Ltd
both of Tiptree, Essex

ISBN 0 09 138091 X

Acknowledgements

In writing this book, I have drawn freely on the writings of research scientists. Any book on livestock production must be based on the work of the late John Hammond and his colleagues at Cambridge. The reader should certainly try to look up Hammond's *Physiology of Farm Animals* and C. P. McMeekan's *Principles of Livestock Production.*

I am extremely grateful to Dr. C. W. R. Spedding for papers on early weaning lambs; J. H. R. Kennedy, N.D.A., Dorset College of Agriculture, for information on Dorset Horn Sheep; Ted Fellows for the use of his excellent shearing drawings, which appear by kind permission of the *Farmer and Stockbreeder*; the British Wool Marketing Board for their illustrations of rolling a fleece; the *Farmer and Stockbreeder* for the photograph of Ashford Rival—Champion Ryeland Ram; H.M.S.O. for the meat production tables from *Annual Reviews and Determination of Guarantees 1970, Cmnd. 4321*; and lastly my wife for providing the roughs from which the illustrations were drawn.

My sincere thanks are especially due to my friends and colleagues Ian Rogerson, A.L.A., County Technical Librarian, Gloucestershire, for his help and encouragement, and for obtaining many scientific papers, etc.; to Harrison Ashforth, N.D.B.; W. W. Mackie, N.D.A., N.D.D., and Stuart Freeman, B.Sc., of Gloucestershire College of Agriculture, for their help and constructive criticism of parts of the text.

I am particularly grateful to my Principal, J. D. Griffiths, J.P., N.D.D., Dip. Agric. (Wales), for the facilities he has put at my disposal, and Gordon Cinderby for reading the proofs.

Lastly I wish to thank Miss Pamela Croft, Mrs. Ruth Smith and Mrs. Janice Wilson for typing the manuscript.

Contents

4 Sheep Farming Systems

5 Sheep Breeds

6 Selection of Breeding Stock for Fat Lamb and Mutton Production

7 The Shepherd's Calendar

13 Wool

14 Handling Pens and Equipment

15 Sheep Housing

16 Common Ailments and Diseases Affecting Sheep

Preface to the First Edition

The aim of this book is to explain in simple language the basic principles of modern sheep production. It is designed primarily for students, past and present, of the county colleges of agriculture, who are either studying or actively engaged in sheep farming.

It is hoped that former students will find much to refresh their memories and a useful guide to present-day techniques. Present students may use the book in conjunction with their course, as background reading, and to reinforce lecture notes. It should also be of general interest to farmers, shepherds, butchers, undergraduates, and others who are connected with the sheep and wool industry.

It must be stressed, however, that a complete mastery of livestock husbandry cannot be gained wholly by reading books or attending lectures. To understand livestock, the student must experience working closely with them. He must be prepared to feed them, clean them, nurse them and observe them at all times, carefully noting the normal and any abnormal behaviour.

All must constantly question farming practice, taking nothing for granted. The questions 'why' and 'how' must be uppermost in our minds:

Why do some ewes lose their fleece shortly after lambing?

Why are ram lambs castrated?

How do you rear an orphan lamb?

Carefully listening, observing and thinking about the answers are sure ways to learning more about a subject.

Preface to the Second Edition

Since this book was first published in 1971 a number of changes have taken place in the sheep industry. This second edition enables me to update much of the previous information and to add topics such as condition scoring; carcass classification; sheep housing and several diseases including sheep scab which was formerly eradicated, but reappeared in 1973.

Over the past seven years the Meat and Livestock Commission's Sheep Improvement Services have provided the farming community with a vast amount of information and sound practical advice. Their research work and market intelligence is proving invaluable to the industry, and I am delighted to be able to use several of their tables to support the text.

My thanks are also expressed to the many farmers, teachers and students, particularly those overseas who have supported my books and made this second edition possible.

Derek H. Goodwin
June 1978

One The Meat Trade

Quality can be best defined as that which the Public like best and for which they are prepared to pay more than average prices.

JOHN HAMMOND
(*Journal of Yorkshire Agricultural Society*, 1955)

Before embarking into any form of business, the wise person will make a careful study of existing and future market. This chapter discusses beef, pork, chicken and meat by-products as well as mutton and lamb, in order that the reader can view an overall picture of the meat trade.

Meat can be defined as 'the flesh or edible parts of domestic animals'; in a broad sense this includes poultry and game. Meat by-products, which are usually referred to as 'offal', include liver, heart, tongue, brains, sweetbreads, kidneys and tripe. Meat and its by-products are sold to the public either in their fresh state or processed as in meat pies, sausage, tinned meat and bacon.

Meat consumption

The consumption of meat is highest in countries with a high national income. In the United Kingdom the present-day (1978) consumption of meat per person per annum is approximately 55 kg. This consists of 21 kg beef, 7·7 kg mutton and lamb, 10·4 kg fresh pigmeat, 10·4 kg bacon, 3·7 kg offal and 11·6 kg poultry (white meat).

Meat is an expensive food to produce and therefore to buy, and beef is dearer than lamb or pork. Poultry, however, is much cheaper today than it was pre-war, a proof of this being the rise in consumption of chicken from 2·25 kg per head in 1938 to nearly 7 kg per head in 1963, and 9·5 kg in 1968.

Meat retailers

The housewife has the choice of purchasing her meat either from a family butcher, or a self-service refrigerated display counter in a grocery store or supermarket.

The family butcher offers a personal service to the shopper, and is always willing to advise his customers on the most suitable joints for their needs. Indeed, the success of the butcher depends largely upon his ability to select and purchase livestock that will yield the type of meat cuts demanded by his customers.

FAMILY BUTCHER

Fig. 1

The supermarkets display pre-packed, boneless joints, with a brief description, the weight and price. The housewife is quite often unable to recognise the cut, and therefore selects her joint solely on appearance and price. However, for the busy housewife, with perhaps a full- or part-time job, the supermarket offers a great saving in time. The pre-packing, self-service food counters are rapidly gaining favour today.

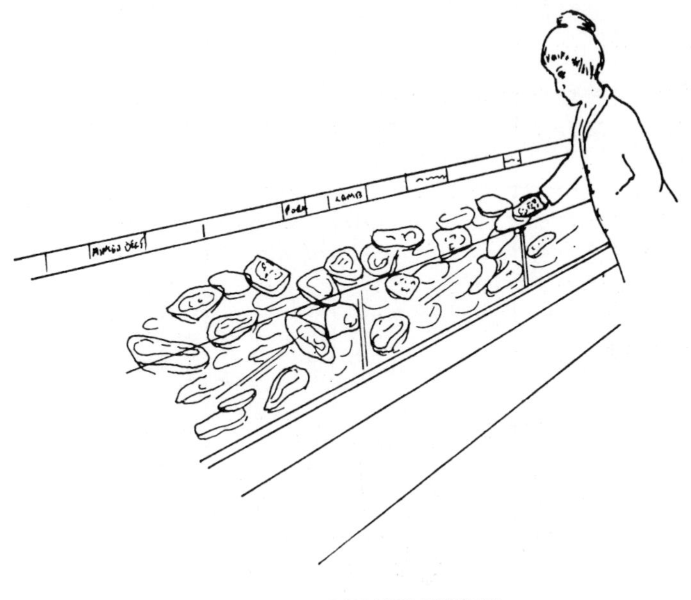

SELF SERVICE COUNTER

Fig. 2

The housewife

The general trend amongst housewives is to buy a small tender joint of meat for the Sunday roast, and to rely on chops, steak, sausage and offals, which can all be prepared quickly, for the mid-week meals. This means that the housewife tends to buy the more 'expensive' parts of the carcass—grilling meat such as steak, chops and gammon are in greater demand than brisket of beef, scrag of mutton, or belly pork. This demand often leads to problems for the retailer in that selling the cheap joints, especially those used for stewing and boiling, is difficult.

There is almost universal demand for lean meat, and quite often the housewife will insist upon lean cuts to the detriment of flavour. Some fat is desirable in meat to aid cooking, but the customer does not always understand this, and will complain when the lean meat with little or no fat from a self-service counter proves to be tough.

Fig. 3 **Livestock numbers in the UK—June** (Figures in thousand head)

	1973	*1974*	*1975*	*1976*	*1977*
Total cattle and calves	14 445	15 213	14 717	14 069	13 925
Total pigs	8 979	8 544	7 532	7 947	7 673
Total sheep	27 943	28 498	28 270	28 265	28 053

Source: MAFF

Fig. 4 **Meat production in the UK** (figures in thousand tonnes) (a)

					January–March	
	1973	*1974*	*1975*	*1976*(c)	*1976*	*1977*
Beef and veal	853·8	1 073·2	1 216·5	1 056·6	293·8	247·0
Pork	681·8	689·5	572·0	584·2	138·0	167·6
Mutton and lamb	234·1	251·5	259·5	242·9	57·1	55·0
Offal	136·1	158·8	166·6	151·8	39·9	36·5
Total (b)	1 905·8	2 173·0	2 214·3	2 035·6	528·9	505·5
Bacon and ham	252·4	243·2	210·2	222·0	52·3	55·3

(a) Including any meat exported
(b) Because of individual rounding, figures do not necessarily add up to totals shown
(c) Converted to standard 52 week year

Source: MAFF

Fig. 5 **Imports of meat into the UK—by country of origin** (figures in thousand tonnes)

		1973	1974	1975	1976	January–March 1976	January–March 1977
Total beef and veal		270·4	249·3	196·4	213·7	37·9	72·2
of which from:	Irish Republic	49·3	104·7	96·4	79·7	12·8	27·7
	Argentina	59·4	28·1	2·4	8·3	3·3	1·0
	Botswana	17·4	3·5	10·7	16·6	3·5	4·6
	France	7·8	14·4	22·8	24·9	4·5	10·0
	W. Germany	3·5	25·6	15·6	23·3	2·6	11·5
Total pork		18·5	7·1	16·3	12·2	1·1	4·2
of which from:	Irish Republic	6·6	1·4	0·4	1·1	..	0·6
	Denmark	6·1	4·8	13·4	10·2	1·0	3·2
Total mutton and lamb		265·7	212·7	243·8	225·7	71·9	80·6
of which from:	New Zealand	239·3	203·5	234·2	213·5	69·5	76·7
	Australia	23·3	7·5	7·6	11·7	2·1	3·9
	Irish Republic	0·9	1·0	1·7	0·4	0·1	0·1
Total bacon and lamb		320·4	298·5	287·4	269·3	67·3	65·7
of which from:	Denmark	248·6	241·4	232·0	208·1	52·0	43·3
	Poland	32·2	19·4	18·4	16·6	4·9	3·3
	Irish Republic	20·8	19·3	9·4	14·0	2·9	5·5
	Netherlands	9·7	12·0	21·1	25·5	6·3	7·7

.. Negligible

Source: Overseas Trade Statistics of the United Kingdom

The housewife also takes particular note of the colour of meat. A light pinky colour indicates freshness, and shows that the joint is cut from a young animal. As an animal gets older, the colour of the muscle darkens (see page 32). Older meat may be tougher, but will have a better flavour than very young meat.

Hanging meat in a coldroom at 4°C for two weeks after slaughter will improve flavour and make the meat more tender—this is due to enzyme action in the carcass. Unfortunately, when meat is hung, the colour is affected and the end product has a rather dark appearance which is less attractive to the customer.

Butchers and meat buyers for supermarkets are very much aware of the housewife's demands for lean and tender meat, and so many buyers purchase only lightweight, quickly grown animals. Thus, the more saleable beef animals are the baby beefs (380 kg–400 kg), which yield a higher proportion of the 'expensive' joints. Similarly with sheep, the butcher prefers to buy young lambs, around 31 kg–36 kg liveweight.

Catering trade

Since the end of the Second World War there has been an enormous change in the eating habits of the British public. Many school children are provided with a cooked midday meal at school. Office and factory workers have canteens, and many business men enjoy expense-account lunches. Restaurants, snack bars, and even 'hot-dog' stands have 'mushroomed' all over the country. There is no doubt that in an affluent society families enjoy eating out regularly, and if the current trend continues we may even see the whole family enjoying their Sunday roast in a competitively priced restaurant. The catering trade can handle the larger joints of meat that is produced from medium to heavy weight stock. Chefs can carve a 3·5 kg leg of lamb that would hardly go into a housewife's oven.

This means that there is a limited market for the heavier sheep of around 38 kg–40 kg liveweight, although it will not command such a high price per kilogram as the lightweight lamb.

Home suppliers

Fig. 3 shows the total supplies of meat in the United Kingdom.

Pre-war we produced about 50 per cent of our meat requirements. By 1959 two-thirds of our needs were produced at home, as they still are today. Yet, despite this tremendous increase in production, the United Kingdom still remains the largest single importing country for meat in the world.

Exporting countries

The four leading exporting countries are Argentina, Australia, New Zealand and Uruguay. The United States of America has a vast home market, and is now importing beef to supplement home production. In Europe, the principal exporters are France, the Netherlands, Denmark, Yugoslavia and Ireland.

Lack of efficiency

As far as technical efficiency is concerned, it is fair to say that the majority of beef and sheep producers are far behind their farming neighbours who produce milk, pigmeat or eggs. Farmers who buy their stock 'by eye' rather than by weight, and feed them on rule of thumb methods, have little idea of the total amount of food consumed, or the food cost per kilogram liveweight gain. Clearly, the sheep producer of the future must use the technical information available to him. Sheep recording (see page 198) in itself will not entirely replace the farmer's eye for good stock, but if the two can be combined, then mutton and lamb can, and must, be produced of consistently high quality, and at a competitive price, with chicken and pork.

It would appear that beef is likely to remain an expensive commodity for some time due mainly to the high cost of production.

Two Reproduction in Sheep

With the exception of the Dorset Horn, all British sheep breeds are season breeders, which means that the ewes will only mate with the ram at certain times of the year. Normally, in the British Isles ewes are mated from August to early December, and in New Zealand ewes will accept the ram from February to August. This is due to the effect of the length of daylight on the ewes, and if sheep are exported from one country to another they will adapt themselves according to their new environment.

It is usual for ewes to come 'on heat' every sixteen to seventeen days. They remain on heat for one to two days when they will accept the ram in the act of mating, generally referred to as 'tupping' or 'service'.

Female reproductive organs

The reproductive glands of all farm animals are very similar. What is discussed here for sheep is basically similar for horses, cattle and pigs.

Ovaries

The ewe has two ovaries which are suspended by ligaments in the loin region of the body and are small rounded bodies, measuring about 12–25 mm across. The ovaries produce eggs or ova and various hormones which assist in reproduction and also affects changes in body growth.

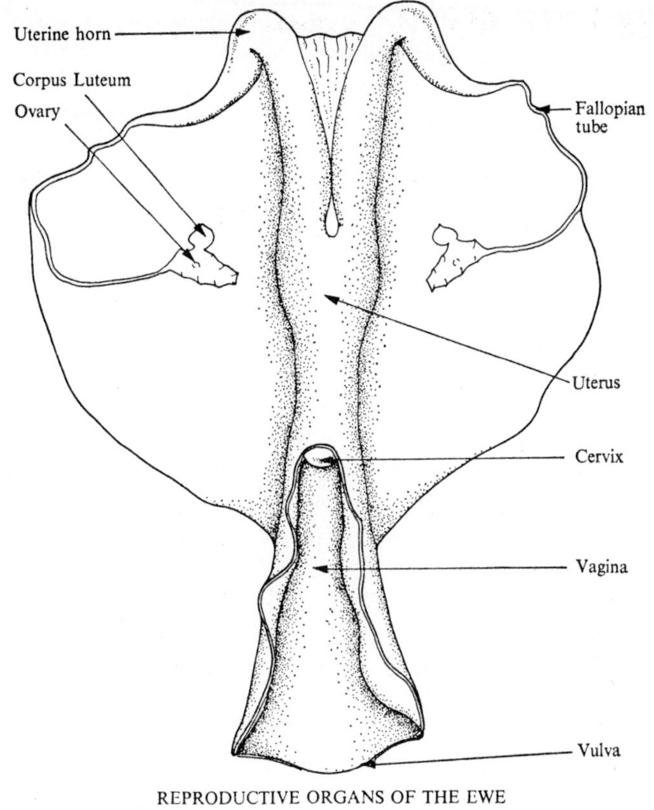

REPRODUCTIVE ORGANS OF THE EWE
Fig. 6

The ova (eggs) are produced in sac-like structures called the Graffian follicles. The follicles also produce a fluid, so when the ovum is fully developed and the follicle ruptures, the fluid carries the ovum from the ovary to the fallopian tube in readiness for fertilisation by the male sperm. This process is known as ovulation and occurs at regular intervals when the ewe is on heat (see page 80). The place of the ruptured follicle is taken by a substance known as the 'yellow body' or 'corpus luteum', which produces a hormone known as progesterone.

If the ewe is mated and the ovum is fertilised, the corpus luteum will persist and produce progesterone which will prevent the ewe coming on heat, and also stop the production of follicles during the pregnancy. If the ewe is not mated, or she fails to conceive, then the corpus luteum will disappear and normal production of follicles will continue.

Fallopian tubes

Linking the ovaries and the uterus together are the fallopian tubes (often referred to as the oviducts because they are used to convey the ova to the uterus) which terminate at the horns of the uterus. Fertilisation of the sperm and ova usually takes place in the fallopian tube, but may be in the uterus.

Uterus

The uterus or womb is the largest part of the reproductive tract. Its function is to provide a home for the fertilised egg until it develops into a fully grown lamb. In sheep this 'growing period' is 147 days and known as the 'gestation'.

Immediately after the ova is fertilised by the male sperm the egg begins to grow and at the same time passes down the fallopian tube into the uterus. The egg attaches itself to the wall of the uterus; a thin 'skin' or 'placenta' is then formed around the growing egg which is now called the embryo. The placenta then produces 'buttons' or cotyledons, which attach themselves to the wall of the uterus, and through which food is passed to the growing lamb via the navel cord or umbilicus.

It can be seen, therefore, that the lamb is not connected directly to the ewe, but is carried and fed inside the placenta.

At birth or parturition the placenta is broken and the lamb delivered without this covering. Later the placenta, now called the 'after birth' or cleansing, is discharged from the uterus and the ewe is said to have cleansed.

Cervix

This is a strong muscular collar that opens and closes the mouth of the uterus. It is normally closed to prevent infection entering the uterus, but is open during oestrous or 'heat period' to allow the sperms to enter, and also at parturition to allow the lamb to be born. The opening and closing mechanism is controlled by various hormones which stimulate the cervix to dilate and contract.

Vagina

The vagina is a fairly large tube which connects the uterus to the external opening called the vulva. In the act of mating the vagina is where the ram deposits his semen; the vagina is also connected to the bladder and used to convey waste urine from the body.

Male reproductive organs

The essential parts of the male's reproductive organs are the testicles, various glands and the penis.

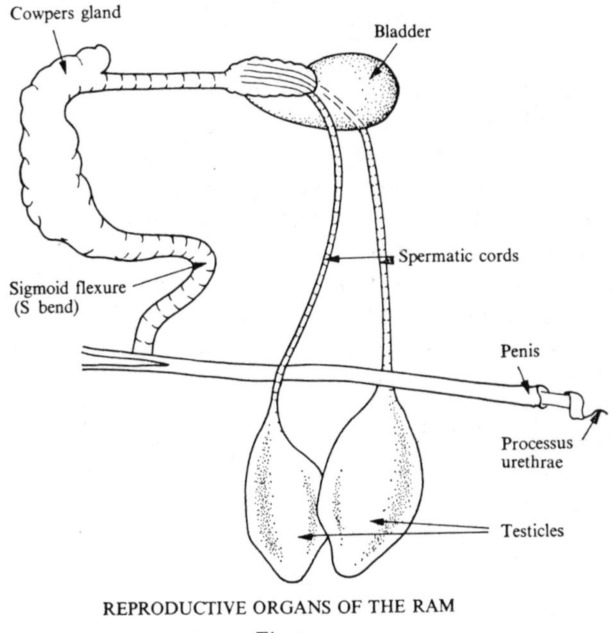

REPRODUCTIVE ORGANS OF THE RAM

Fig. 7

Testicles

The ram, like other male farm animals, has two testicles, which are suspended outside the body, but enclosed in a sac or scrotum. The normal temperature of the testicles is below that of the body; the scrotum is outside the body in order to regulate the temperature. Each testicle is responsible for producing large amounts of live sperms; these are stored temporarily in the epididymis—a small knob-like structure situated at the base of the testicle.

It will be observed that a ram has much larger testicles in relation to its body size than a bull. This is to enable the ram to serve a large number of ewes in a short time.

The sperms are carried in fluid from the epididymis, along tubes to the vesiculae seminales, where they are mixed with other secretions from various glands, before passing into the uretha for ejaculation.

The penis is of considerable length in the ram, and bends to form an S shape known as the sigmoid flexure. In the act of mating the S is straightened, thus allowing the penis to extend and enter the ewe's vagina.

Three Growth and Development of Sheep

Hammond defined growth as 'the increase in weight of an animal, until a mature size is reached'. Development is defined as 'the change in body shape or conformation'. Development also includes changes in body structure, for example, development of mammary tissue in the pregnant female.

Growth and development before birth

Life begins at conception, with the union of the male sperm and the female ovum; each parent contributes substantially one-half to the inheritance of the offspring. In the early stages of pregnancy the developing foetus is surrounded by large amounts of fluids and tissue. In the later stages of pregnancy the unborn animal makes rapid growth in size and weight, its bulk replacing much of the fluids.

It is important, therefore, that all pregnant stock should receive adequate nutrition during the final stages of gestation in order that the developing animal is sufficiently well grown and is strong and active at birth. Supplementary feeding of ewes before lambing is known amongst farmers as 'steaming up'. A useful 'yardstick' is to remember that two-thirds of growth takes place in the last third of the gestation.

Birth weight

The size of the offspring at birth is controlled by factors other than nutrition during pregnancy. The more important of these are sex, breed and single or multiple births.

Thus, in sheep, Southdown lambs are smaller than Suffolk lambs; twins and triplets are generally smaller at birth than singles but their total weight may amount to more.

Size at birth is also controlled by the dam, by means of a special substance in the mother's blood-stream. This substance prevents the foetus from growing to such a size as to prevent parturition. Were it not for this control, the mating of a large breed like the Border Leicester to a small breed like the Welsh Mountain would give rise to difficulties at birth. These occur only in exceptional circumstances. Generally, lambs born from hoggets and yearlings are lighter in weight than from older ewes. Overfat ewes tend to produce lighter lambs than ewes in a good fit condition.

The birth weight is important because it influences subsequent growth rate. Strong healthy lambs that are heavy at birth grow more quickly during the first two to three months, for example a 7 kg single lamb will reach around 36 kg in ten to twelve weeks, if tended well, whereas a similar lamb weighing only 4 kg–4·5 kg will take about two weeks longer. Small lambs, however, will eventually catch up, and providing their small size is not due to heredity, will become just as heavy at maturity as lambs that were heavier at birth.

Growth after birth

After birth, growth is usually measured as kg daily liveweight gain (d.l.w.g.), or weekly liveweight gain (w.l.w.g.).

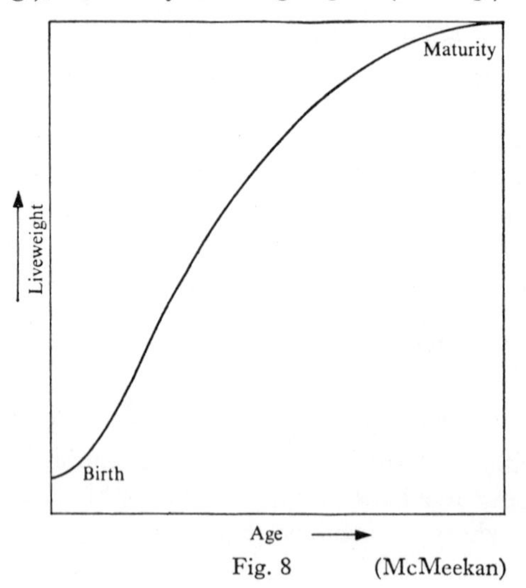

Fig. 8 (McMeekan)

In all animals the growth rate is slow to begin with, and then rises quite rapidly but slows again as maturity is approached.

If the d.l.w.g. is plotted against the age of the animal on a graph, then we get a characteristic S curve.

The S curve is similar in all farm livestock, but the degree of steepness in the curve will vary according to the breed and strain of animal.

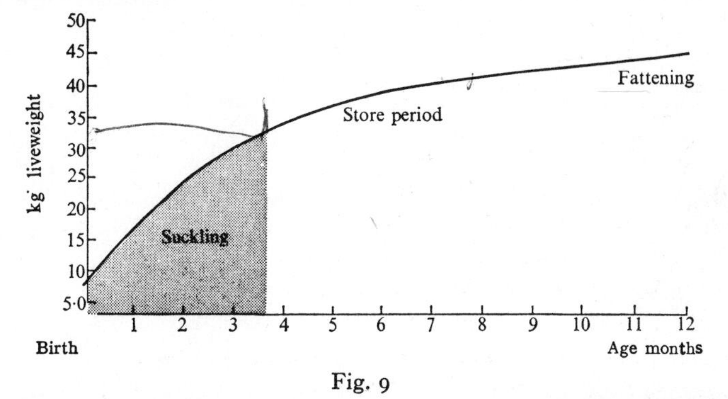

Fig. 9

Lambs grow very rapidly indeed during the first few months of their lives, and many in fact reach slaughter weight in less than half the time that they were in the uterus! (Gestation period twenty-one weeks.) On lowland farms using Down rams on Half-bred ewes, lambs may be expected to put on at least 9·0 kg liveweight in the first month, 7 kg in the second, and 6 kg–6·5 kg in the third. Thus lambs weighing 6 kg at birth will easily reach 26 kg or more in twelve weeks if managed well. After weaning lambs usually pass through a 'store' period, which means they grow rather slowly; the 'store' period is often associated with a drop in the level of nutrition—pastures are less productive in July and early August. Lambs that are fattened in the autumn with the aid of supplementary concentrates will gain about 1 kg per week.

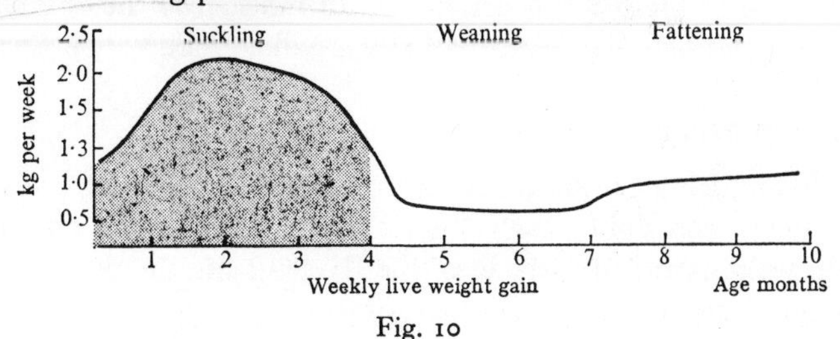

Fig. 10

Changes in body proportions (waves of growth)

At birth the head is relatively large, the legs are long, and the body small. In the mature or fully grown sheep the head is small and the legs short in proportion to the body.

Changes in body shape are due to different parts growing at different rates. The head and legs are early to develop (skeleton), while the body, particularly the hindquarters and loin region, are late in developing and are the last parts to reach mature size.

The body, which is later to become the carcass, consists of three main tissues, *bone* which forms the skeleton, *muscle* which is the lean meat, and *fat*. These three tissues grow in a very definite order. Bone is the first to develop, muscle is intermediate but tends to follow bone fairly closely, while fat is the last to develop and grows fastest as the animal approaches maturity.

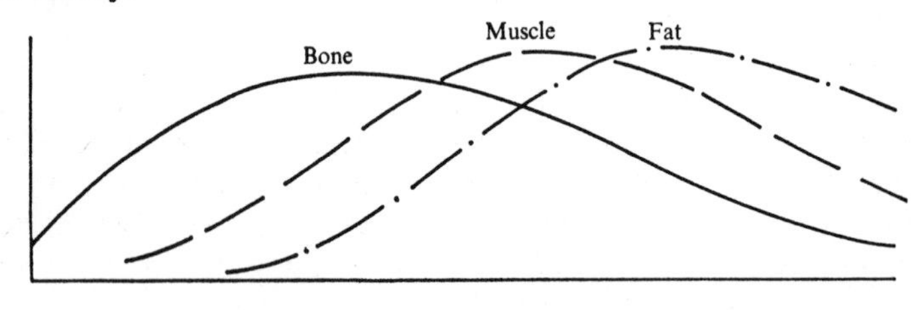

WAVES OF GROWTH OF THE MAIN TISSUES

Fig. 11

The young animal, therefore, contains a higher proportion of bone and muscle and a lower percentage of fat than the older animal. This knowledge is of vital importance to meat producers, since the present-day demand is for young, tender, lean meat. It can readily be appreciated that if livestock are fed on a high plane of nutrition during their early life, then the final carcass will have the maximum proportion of muscle. If the animal is slaughtered before it nears maturity it will contain very little fat.

Early maturity

Early maturing breeds are those in which the 'waves of growth' are steep, and follow each other closely. Such breeds fatten at light weights and at early ages: the Ryeland and the Southdown are good examples of early maturity.

Late maturity

Late-maturing breeds, like the Lincoln and Leicester, may grow fast in terms of d.l.w.g., but because their waves of growth are wider apart than early maturing stock they take longer to fatten and mature. One of the problems with late-maturing breeds is that lambs will reach the popular market weight (31 kg–36 kg) quite quickly but lack sufficient 'finish' to be suitable for slaughter (see page 119 grading lambs).

Changes in internal organs

In addition to the changes that take place in the body, similar 'waves of growth' occur in the internal organs. The main body organs develop in the following pattern: first the brain, secondly the heart and thirdly the digestive system. The reproductive organs and the udder tissue in the female do not develop until much later.

In sheep the lamb has a well-developed brain at birth, its lungs are well developed and so is the heart, but the digestive system does not fully develop until the lamb is three weeks old. The reproductive organs start to function at puberty, when the lambs are 6–8 months old. The udder develops very little until the ewe is in-lamb and continues to develop during the subsequent lactations.

Offals and dressing percentage

When the animal is slaughtered most of the internal organs are utilised as offal (see page 119).

Young animals will have a higher proportion of offals in relation to carcass weight than will the older animal. This explains why young stock have a lower dressing percentage than older stock.

		Dressing
Suckling lambs	12–16 weeks	48–50%
Fat ewes	3–4 years	56–58%

Evaluation of the carcass

Muscle tissue

Muscles are made up of fibres, arranged in bundles. As the

fibre grows, the muscle bundle enlarges, which causes the meat texture to become coarse and less tender. Some parts of the carcass, such as the loin and leg, are more tender and have a finer grain than the other parts in the same carcass, such as the neck and shoulder.

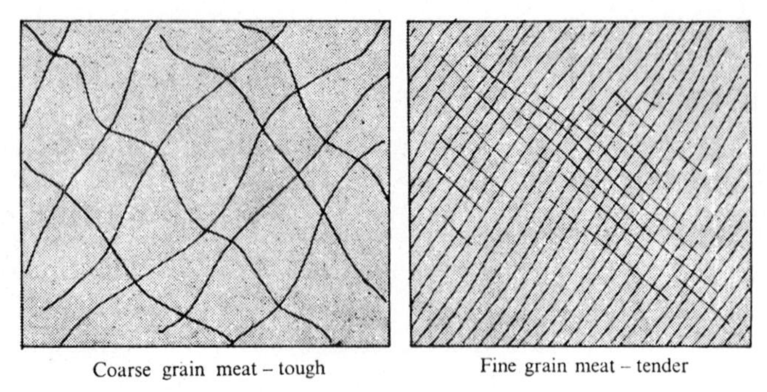

Coarse grain meat – tough　　Fine grain meat – tender

Fig. 12

Colour of muscle

The colour of muscle in the young animal is always lighter than that of older stock; this is due to a pigment that is released in the movement of the animal. Thus, the more exercise that the animal takes during its lifetime, the darker the colour of the muscle. Exercise also strengthens the muscle fibres, and this makes the grain become coarse and the meat tough.

Veal calves produce light-coloured carcasses. Pork is always lighter than beef or mutton.

Flavour is also associated with the same pigment that affects the colour of the meat. Thus, in the older animal (e.g. mutton), the flavour is more pronounced, but the meat tougher to eat. Likewise, very young meat such as spring lamb, broiler chicken, veal and pork tend to lack flavour, although, of course, they are extremely tender.

Fat

During the lamb's early life fat is deposited first on the gut and kidney regions, then a layer over the muscle (this gives a smooth, firm appearance to the well-finished lamb), and, lastly, the fat penetrates the muscle bundles and causes 'marbling', a term that butchers use to describe intramuscular fat. Marbling is desirable in older animals such as fat ewes, because it breaks up the muscle fibres, making the meat more tender.

Although overfatness is not usually a problem with suckling lambs, hoggets and ewes tend to become overfat if not carefully managed. There is a marked reluctance by housewives to buy ewe mutton that is overfat.

Summary

Growth is the increase in weight in an animal until it reaches maturity, and development refers to changes in body shape.

The unborn lamb grows slowly for the first 100 days or so of the pregnancy and then makes rapid growth during the final 6–8 weeks. It is important, therefore, that ewes are 'steamed up' to get strong lambs at birth and ewes with plenty of milk.

Lambs grow quickly during the first 10–12 weeks (1·8 kg–2·25 kg per week) provided they receive adequate milk from the ewes and are grazed on clean pastures. After weaning they grow rather slowly (0·45 kg–1 kg per week) through their store period and finally put on about 1 kg–1·3 kg per week in the fattening period.

Four Sheep Farming Systems

The aim of this chapter is to given an overall view of the sheep industry in Great Britain. A brief description of the management of a hill flock is given; stratification is explained and the breeding of crossbred ewes is mentioned; figures are kept to a minimum, since they quickly date, and are used only to illustrate the pattern of farming and not to make a study of sheep economics. The reader should gain this background knowledge of the industry before proceeding to subsequent chapters which deal in detail with the day-to-day management of the lowland flock.

The broad pattern of sheep farming is closely related to the beef industry since beef and sheep are complementary to one another and rarely competitive. Sheep are hardier than cattle; they are able to thrive at higher altitudes and withstand both high rainfall and drought better than beef animals.

The main purpose of sheep husbandry in the United Kingdom is the production of meat, i.e. fat lamb and mutton, most of which is produced on lowland farms. Wool is of secondary economic importance, though it is essential for hill breeds to grow strong fleece as a protection against bad weather. In lowland flocks the income from fat lamb is about five to six times that of wool, whilst in hill flocks wool accounts for almost half the income.

Stratification

The main production of commercial fat lamb and mutton is maintained by an integrated cross-breeding system which we call stratification. This involves the drafting of commercial

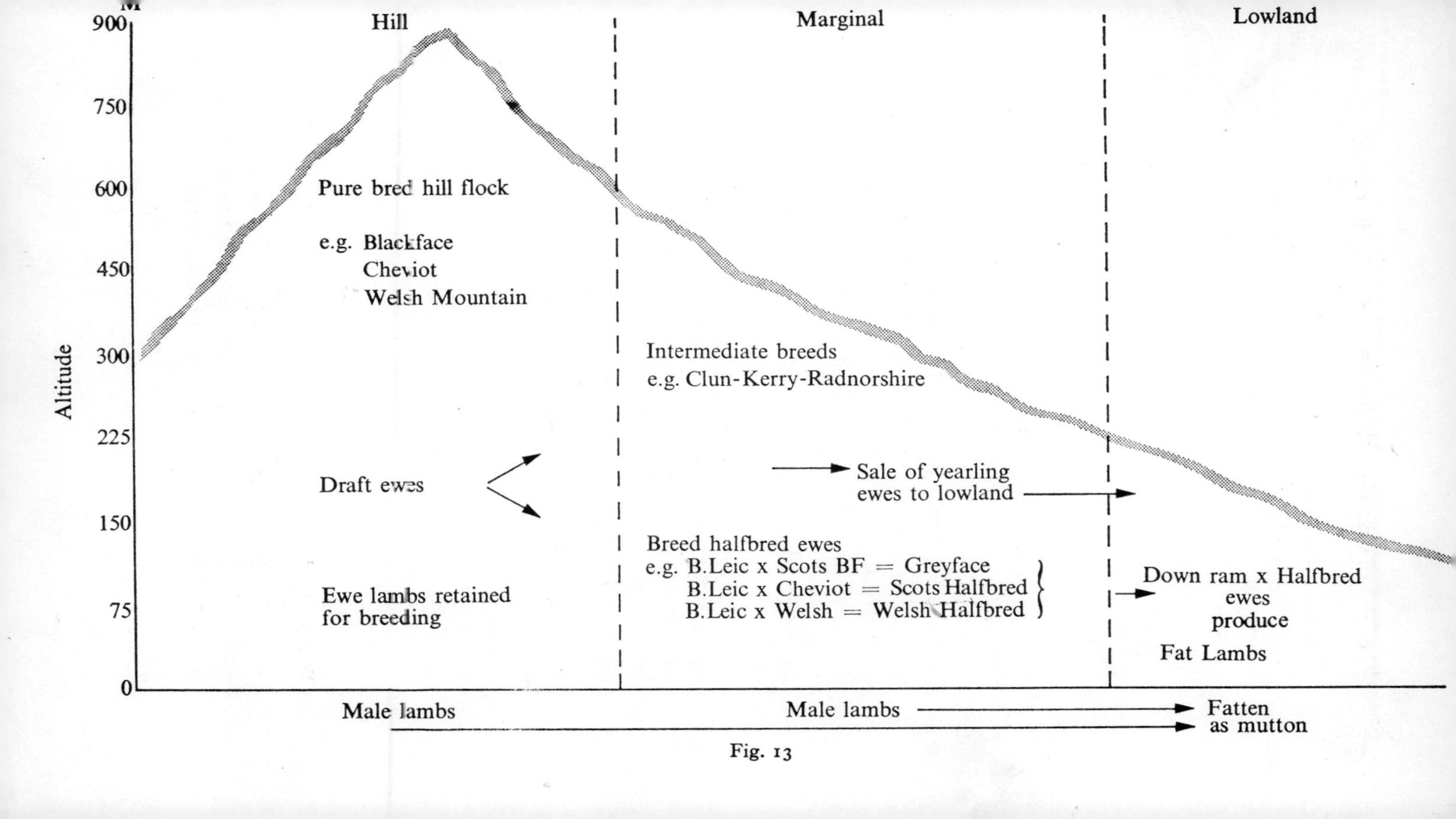
M
900
750
600
450
300
225
150
75
0
Altitude
Hill
Marginal
Lowland
Pure bred hill flock
e.g. Blackface
Cheviot
Welsh Mountain
Intermediate breeds
e.g. Clun-Kerry-Radnorshire
Draft ewes
Sale of yearling
ewes to lowland
Breed halfbred ewes
e.g. B.Leic x Scots BF = Greyface
B.Leic x Cheviot = Scots Halfbred
B.Leic x Welsh = Welsh Halfbred
Ewe lambs retained
for breeding
Down ram x Halfbred
ewes
produce
Fat Lambs
Male lambs
Male lambs
Fatten
as mutton

Fig. 13

breeding ewes from the mountains and uplands to the lowlands for crossing with Down rams (see Fig. 13).

A good example of stratification is the production of Scots Half-bred ewes. Cheviot ewes from the mountains, after they have produced two or three crops of lambs, are sold as draft ewes to farmers in the uplands and there mated to a Border Leicester ram. The resulting ewe lambs are kept on the upland farm until they are fifteen months old and are then sold as yearling Scots Half-breds to farmers in the lowlands. The Scots Half-bred ewe is then mated to a Down ram, such as the Suffolk, to produce first-quality commercial fat lambs.

There are many advantages in cross-breeding sheep. Firstly, cross-breeding offers hybrid vigour, which is the superior performance of the progency over the average performance of the parental breeds. Cross-breeding makes it possible to combine qualities of two or more breeds. For example: the mothering instinct, hardiness, and milking ability of hill breeds are fused with the prolificacy and relative early maturity of the longwool sheep. Thus, cross-bred ewes such as the Scots Half-bred, Welsh Half-bred, Masham and Greyface, are deep milkers, hardy, prolific, quick maturing and possess size and scale: or, simply, are just 'good ewes'.

To illustrate the importance of cross-bred ewes it is worth studying the annual livestock statistics. The 1977 figures show the total sheep population in the United Kingdom to be 28 million sheep, including:

11 million breeding ewes. The ewe flock is further divided as follows:
4¾ million hill ewes—mostly pure bred
4½ million cross-bred ewes producing further cross-bred ewe lambs and fat lambs
¾ million draft hill ewes, producing cross-bred ewes
¾ million pure-bred ewes not registered in flock books
½ million pedigree ewes, mainly used for producing rams for crossing.

The hill farm

Approximately 25 per cent of the land in the British Isles is classified as hill grazing, ranging from the rugged heather-clad

Fig. 14 **Sheep numbers in the United Kingdom**

June	*1973*	*1974*	*1975*	*1976*
Ewes for breeding	10 921	11 192	11 279	11 298
Shearlings to be put to the ram in the current year	2 733	2 673	2 471	2 369
Rams for service	314	322	326	320
Other sheep and lambs:				
1-year-old and over	910	947	952	829
under-1-year-old	13 066	13 364	13 222	13 449
Total sheep and lambs	27 943	28 498	28 270	28 265
December				
Ewes for breeding	12 840	13 018	12 746	12 709
Shearlings put to the ram				
Ewe lambs put to the ram	788	601	576	708
Ewe lambs for breeding but not put to the ram	1 920	1 836	1 789	1 883
Rams for service	343	357	344	343
Other sheep and lambs:				
1-year-old and over	791	802	731	726
under 1-year-old	3 511	3 571	3 349	3 529
Total sheep and lambs	20 193	20 187	19 536	19 899

(a) Provisional

Source: MLC

Fig. 15 **Distribution of breeding ewes (%)**

	Hill flocks	*Upland flocks*	*Lowland flocks*	*All flocks*
England	10·2	5·5	29·6	45·3
Wales	14·1	6·4	5·5	26·0
Scotland	20·0	5·3	3·4	28·7
Great Britain	44·3	17·2	38·5	100

Fig. 16 **Distribution of breeding ewes**

Type of Farm	%
Livestock rearing	59
Dairy	12
Mixed and cropping	19
Others	10

Percentage of farms that have sheep

31% in England
50% in Wales
50% in Scotland

Average flock size

England	147 ewes
Wales	203 ewes
Scotland	213 ewes

Percentage of flocks above 500 ewes

28% in England
30% in Wales
52% in Scotland

Source: MLC

mountains in the Scottish Highlands, over the bleak Cheviot Hills bordering England and Scotland, through the Lake District, the Pennines, the mountains of North and Mid-Wales, to the peats, bogs and open heaths as far south as Exmoor and Dartmoor.

It is almost impossible to select a typical hill farm, but suffice it to say that hill farms are extensive rather than intensive. Much hill land is extremely poor, usually acid and capable of growing only the poorest species of grass. Often it is necessary for eight or more hectares of hill land to support a ewe and lamb for grazing.

Scottish hill farms are larger than Welsh farms. A Scottish farm may extend to 1000 or 5000 ha, whilst a Welsh farmer often gleans a living from as little as 100 to 150 ha of open hill, with an inbye of some 20 hectares.

Breeding stock

Under true hill-farming conditions a pure-bred flock is kept. The sheep become acclimatised to the local conditions: soil, climate and pastures; hence the flock is said to be 'bound to the ground'. If the farmer decides to leave his farm he will hand the flock over to the next occupier at current valuation.

Income

The hill farmer's main income is derived from the sale of draft ewes, wether lambs, and wool. The lambing percentage in hill flocks is unlikely to exceed 80 to 100 per cent if taken as an average over several years (i.e. 80 to 100 lambs from 100 ewes). In mild winters 120 lambs may be reared per 100 ewes put to the ram, although too high a proportion of 'twins' are not favoured amongst hill flock masters. Single lambs are stronger at birth and will thrive on the open hill. Twin lambs, however, often go short of ewes' milk, and become stunted if turned to the hill.

The farm

A hill sheep farm consists of two well-defined areas: the inbye, which is the fenced area, containing the farmhouse, buildings and sheep pens, and the higher open hill, which may be partly fenced, or left completely open.

The inbye holds the ewe flock during the severe winter months until early spring, when lambing takes place. The inbye is used periodically in the summer months for gathering the flock for dipping, shearing and routine management tasks. The inbye, cultivated by the farmer, grows grass for conservation either as hay or silage, and sometimes a few acres of oats are also grown. A small acreage of grass is ploughed out from time to time for reseeding, when a break crop of swedes, turnips, or rape and kale may be sown, or the field directly reseeded to grass.

The hill

The herbage on the hill consists of the poorer-quality species of grass such as bents, molinia, cotton grass, Yorkshire fog and fescues, together with a wide range of herbs, weeds, sedges, rush and shrubs, such as heather and gorse. Heather is particularly valuable, owing to its ability to stand above the early light falls of snow, offering feed for the sheep from its young shoots. In most hill areas a small proportion of the older heather is 'burnt off' each year in order to encourage new growth, which, of course, is more palatable and nutritious for the sheep.

Routine management

On the majority of hill farms in Wales the flock is shepherded by the farmer, with occasional assistance at busy times from his neighbours. In Scotland, where farms tend to be larger, a shepherd is employed to look after some 600 to 1,000 or more ewes. His main duties will be walking the hill each day to inspect the ewes, and to treat any ailments. He may ride a pony when the flock is some distance from his home, and, of course, he will always be helped by two or more trained sheep-dogs.

We will now look briefly at the shepherd's work throughout the year on a hill farm. The mating season, commonly referred to by the shepherds as 'tupping' time, starts on the hill farm in November and lasts through December. Owing to the extreme weather conditions in winter and early spring lambs are not required until early April to May. Traditionally, shepherds turn out the rams in Scots Blackface flocks on 25 November, whilst Cheviot breeders put their rams with the ewes on 10 November.

One yearling or mature ram is allowed for every 30–40 ewes; ram lambs are not used with hill flocks. Flushing (see page 79) is not practised in hill flocks since the aim is to breed strong, vigorous, single lambs which have a much better chance of survival than weakly twin lambs, and flushing tends to make the ewes more prolific.

Lambing time

In fairly sheltered areas the flock may stay out on the hill all the winter, and be brought down to the inbye only at lambing time, but on most hill farms the flock is brought to the lower part of the hill or to the inbye as soon as the first falls of snow arrive.

Traditionally, the ewes are fed hay and roots only, but in recent years more and more farmers are feeding a small amount of concentrates—say 0·25 kg–0·5 kg per head per day—during the last stages of pregnancy and during the first weeks of suckling.

By mid-May, the hill will be 'greening up', which means that the growing season has commenced, and the flock is then returned to the hill for the summer. Before this, however, all the ram lambs are castrated and docked. The ewe lambs will only have the short tip of their tails removed, as the tail is left to protect the udder. All the tails and tips are placed in a bag, and at the end of the day the farmer counts the tails to know how many lambs are being turned to the hill.

Every farmer has a paint brand with which he marks his sheep; some farmers also ear mark their lambs with a tattoo or ear-tag (See Fig. 17). Horned breeds may be horn branded.

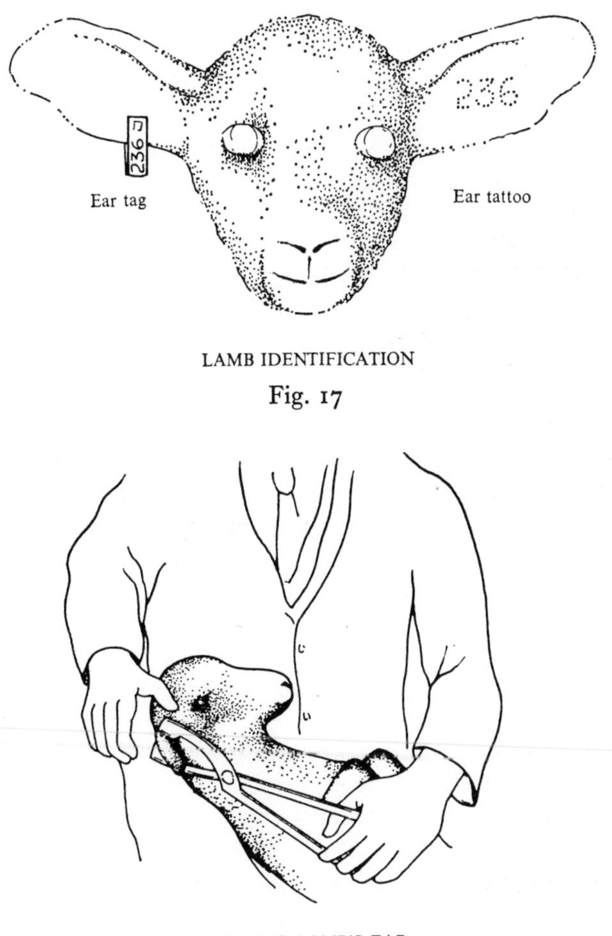

LAMB IDENTIFICATION

Fig. 17

TATTOOING LAMB'S EAR

Fig. 18

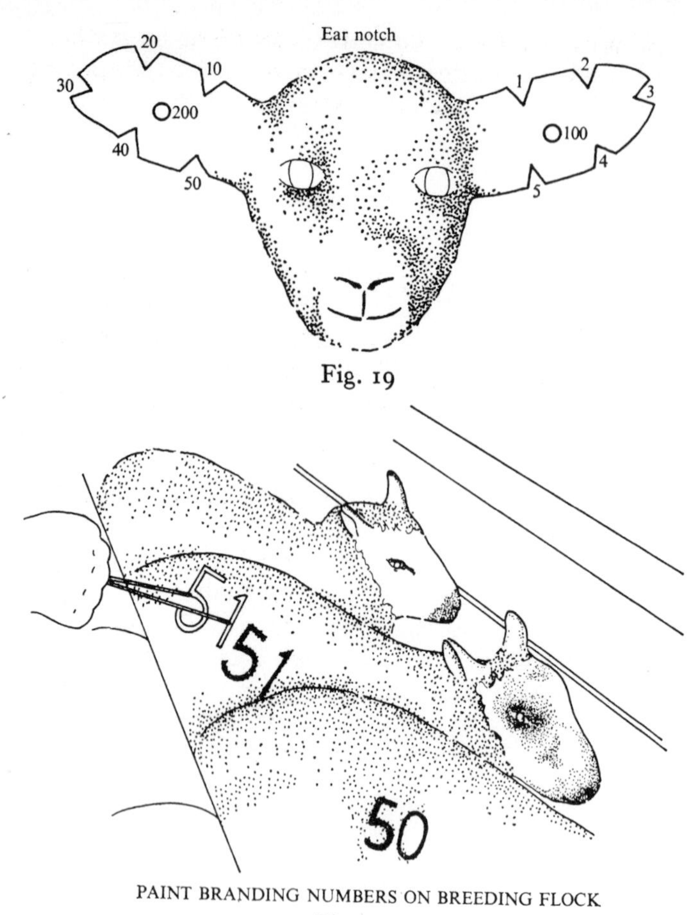

Fig. 19

PAINT BRANDING NUMBERS ON BREEDING FLOCK

Fig. 20

Summer months

The flock remains on the hill throughout the summer months, and is only brought down to the inbye for shearing and dipping. Many hill flocks are still clipped by hand because the hand shears leave a small amount of wool on the sheep to give it immediate protection from heavy rain and cold nights. There are, however, special sheep-shearing combs that leave some wool on the skin, and these are now being more widely used.

Shearing takes place in July and August when the weather is warmest, and is usually undertaken by several shepherds, working as a team, and moving from farm to farm.

Autumn sales

The hill farmers' main source of income is the sale of draft ewes and castrated male lambs, which are sold in September and early October. Old ewes that are sound in feet, teeth and udder, but are unsuitable for keeping on the exposed hill, are sold to farmers in the uplands and sometimes the lowlands as draft ewes. Usually these hill ewes are mated to a longwool ram to produce a 'half-bred' ewe lamb (see page 36). The wether lambs are sold to the lowlands to be fattened on roots, sugar-beet tops, or grass, hay and corn.

The ewe lambs are retained for breeding, but, unlike the lowland ewe lamb, they are not bred from until they are 15–18 months old or yearlings. Many farmers send their ewe lambs down country for the winter months. This is referred to as 'tacking out lambs'. The hill farmer pays so much per head (in recent years £3–£4 per head) to the lowland farmer who feeds the lambs until the following spring.

Winter housing

There is now considerable interest amongst hill farmers in wintering their ewe lambs in buildings and buying in extra food. The cost of housing and feeding is rather less than tacking the lambs; there are now several farmers who are keeping their ewes indoors during the winter. Under true hill conditions there is an enormous difference between the growth of herbage during the short summer season and the almost semi-starvation period during the long winter months. In these conditions, there is no doubt that winter housing the breeding stock can greatly improve the performance of the flock. The cost of housing could be partly offset by fattening the wether lambs indoors during the autumn and selling the final product direct to the meat buyer.

The upland marginal sheep farm

The upland farms, with their improved pastures, better buildings, and better amenities than the hill farm, offer almost ideal conditions for sheep breeding. The border counties of Montgomeryshire, Radnorshire, Breconshire and Monmouthshire are good examples of the uplands. Here the pure-bred intermediate breeds like the Clun, Kerry Hill, Radnorshire and Beulah Speckled Face are bred alongside the pure-bred Welsh Mountain.

The flocks lamb down in March to early April, which is slightly ahead of the grazing season. The lambs make rapid growth during the summer months as they are said 'to grow with the grass'. Quite often, wether lambs are sold straight from their dams to the butcher at 30 kg–36 kg liveweight. The ewe lambs are retained on the farm until they are fifteen months old and then sold as yearling ewes at the border county ewe sales.

The lowland sheep farm

There is a wide variety of sheep-farming systems to be found in the lowlands, although it is fair to say that the lowland farmer is the 'finisher' of fat lamb and mutton.

A small number of farmers specialise in breeding pure-bred Down rams for sireing commercial fat lambs; other farmers buy store lambs in the autumn to fatten as mutton. The majority of sheep farmers, however, keep either half-bred or intermediate ewes to produce fat lambs off grass. Chapters Six and Seven deal with lowland sheep production, but three slightly unorthodox systems are mentioned here.

Out-of-season lambing

A small number of farmers produce out-of-season lambs by keeping a flock of Dorset horns and Polled Dorsets. These ewes are capable of breeding at any month of the year, but in practice they are managed so as to lamb in late February–March and again in August–September. A further possibility is to lamb part of the flock in July, part in December, and part in April, thus the ewes lamb at eight-monthly intervals and raise three crops of lambs in two years. Both systems call for a high standard of shepherding, and strict discipline regarding putting the rams out with the ewes, and removing them at the correct time, otherwise the system will quickly deteriorate into one continuous lambing process. Out-of-season lamb commands a premium price, especially around Christmas time.

Flying flocks

These consist of ewes that are purchased in the autumn and kept until the following spring or early summer, when the entire flock is sold off the farm. This system has several advantages: firstly, the farmer has a quick turnover of capital, especially if he sells the ewes with lambs at foot as 'couples' in the early spring; secondly, the farm is rested every year from sheep during the late summer months, and this helps to control contagious diseases such as foot rot and parasitic roundworms. Buying in fresh ewes each year allows the sheep a 'change of ground', one of the secrets of successful shepherding.

Farmers who keep flying flocks usually select the older ewe, of say 4–5 years (full mouth); these are relatively inexpensive to purchase and are generally more prolific than young ewes. If any ewe fails to breed she is sold as a fat ewe in January or February when there are good prices for ewes. If the flock lambs down early in the year, then the farmer will be able to fatten the ewes after the lambs are sold.

All grassland farms on the Romney Marsh

The Romney Marsh in Kent provides some of the finest pastures in the world. Over the past 200 years a highly specialised system of grazing has been developed, whereby graziers practise stocking rates of around twenty wether sheep per hectare. The land is rested during the winter, yet many farmers claim that they never fertilise or mow the pastures! Furthermore, the Romney sheep rarely seem to be troubled with foot rot.

References

H.M.S.O., *The Structure of Agriculture*, Ministry of Agriculture, Fisheries and Food, Department of Agriculture, Scotland.

McCOOPER, *Taking Sheep off the Hills*, Agriculture Vol. 70, No. 10.

Five Sheep Breeds

There are over forty sheep breeds of commercial importance kept in the United Kingdom, and they are classified as Mountain, Longwool, Intermediate and Shortwool.

Mountain breeds

The mountain breeds (found on mountain, hill and moorland at 300–900 m) are generally small sheep, alert in appearance, very hardy and able to live and thrive in exposed areas. Some breeds are horned, and the face colour varies from white to black. However, all mountain breeds are alike in their ability to produce excellent quality mutton when fattened. The joints are small, of fine texture, and free of excess fat, the carcass being light in bone. The wool is extremely strong and durable though poorer in quality than lowland breeds, but one must remember that the biological value of wool is to protect the sheep from the cold, rain and snow. The shorn fleece is used for manufacturing tweeds, blankets and carpets (see page 151, wool quality and spinning counts).

Longwool breeds

The longwool breeds are usually associated with rich fertile lands where cheap arable by-products and an abundance of grass are found. The earliest improvement in longwool sheep was made by Robert Bakewell (1725–95) by line-breeding and in-breeding of Leicester sheep.

Longwool breeds are all free from horns, white-faced, big-bodied, often weighing 90 kg–130 kg liveweight, and may carry fleeces weighing upwards of 9 kg. The wool is long, coarse, and lustrous and varies in quality according to the breed. The chief longwool breeds are the Leicester, Border Leicester, Teeswater, Wensleydale and Romney Marsh, whilst South Devon and Devon Longwool are particularly popular in their native county. Longwool breeds are later to mature than shortwool breeds and tend to produce joints too large for present-day requirements.

Shortwool breeds (including Down breeds)

The shortwool breeds are spread widely through the lowlands, particularly in the western counties of England. The Down breeds are hornless, dark-faced, dark-legged and carry close, fine fleeces of good quality, and also are supreme as mutton and fat lamb producers. Stock rams of Down breeding are used for crossing with upland and half-bred ewes to produce commercial fat lambs. The Down breeds are the Southdown, Suffolk, Hampshire, Oxford, Shropshire and Dorset Down.

The other shortwool breeds are the Ryeland, a breed similar to the Southdown, although somewhat larger; the Wiltshire horn, white-faced and without wool; the Dorset Horn—a breed capable of producing two crops of lambs per year; and the Devon Closewool—a popular breed in Devon, Dorset and Cornwall.

Intermediate breeds

The intermediate breeds, Clun, Kerry Hill, Radnorshire and Speckled Face, all originate from mountain breeds. They are found in the border counties of England and Wales. Many flocks are kept at altitudes in excess of 300 m above sea-level, but in general the intermediate classes are kept in marginal and upland areas as pure-bred flocks producing yearling ewes for sale to the lowlands.

Well-advertised sales of these popular breeds are held at Craven Arms, Knighton, Hay-on-Wye, Kington and Builth Wells.

Rare breeds

Over the past thirty years several of our native breeds of domesticated livestock have become extinct. This has caused certain farmers, scientists, and conservationists to set up various ways of preserving breeds which are likely to be lost. In 1974 The Rare Breeds Survival Trust was formed with the aim of maintaining breeds of animals that were dangerously low in number. The Trust has been very successful, and today there is considerable interest in keeping rare breeds, especially sheep. Breeds like the Jacob and Black Welsh Mountain are already well on the increase, and can be seen at many agricultural shows. Other breeds like the Portland and Manz Loghtan are still very low in numbers, and usually only found at one of the Trust's farm parks.

Classification of breeds

Mountain
Scots Blackface
Swaledale
Welsh Mountain
Black Welsh Mountain
Cheviot
North Country Cheviot
Herdwick
Exmoor Horn
Rough Fell
Lonk
Derby Gritstone
Penistone
Whitefaced Dartmoor

Longwool
Leicester
Border Leicester
Blue-faced Leicester
Lincoln Longwool
Wensleydale
Romney Marsh
Devon Longwool
South Devon
Teeswater
Cotswold

Shortwool—not Down breeds
Ryeland
Wiltshire
Dorset Horn
Devon Closewool

Intermediate
Clun Forest
Kerry Hill
Hill Radnor
Beulah Speckled Face
Llanwenog

Down
Suffolk
Southdown
Oxford
Hampshire
Shropshire
Dorset Down

Rare breeds—unimproved
St Kilda
Manz Loghtan
North Ronaldsay or Orkney
Portland
Whitefaced Woodland
Soay
Jacob
Shetland

Imported breeds
Friesland
Texel
Île de France
Oldenburgh
Finnish Landrace

The following is but a brief description of the more important breeds found in the British Isles. In selecting a breed the newcomer to sheep husbandry is strongly recommended to visit flock masters and discuss the merits of the breed he is interested in. Most breeds, but not all, have a Flock Book Society which publishes literature, organises shows, sales, etc., and generally furthers the interests of their breed.

Mountain breeds

Scots Blackface

This is, undoubtedly, the most important breed of sheep in Scotland and probably in the British Isles. It is easily recognised by its black face with white spots, and both the male and female are horned. The body is small in size and covered with 1·8 kg–2·25 kg of medium-quality wool, which is used for making carpets and tweeds.

The breed is famous for supplying draft ewes to the uplands, which, when crossed with the Border Leicester ram, produces the well-known Greyfaced ewes. Store lambs are sold to the lowlands to be fattened as quality mutton—small, lean joints of fine-grained muscle, free from waste fat and relatively light in the bone.

Cheviot

This is a white-faced breed which is said to be related to the Welsh Mountain: certainly, when comparing the wool with the Welsh, there is a marked similarity.

Mountain flocks clip 1·3 kg–1·8 kg, and hill flocks slightly more. The wool is better quality than the Scots Blackface, but not as good as the Down breeds. The body is small but long, and generally dips in the back.

The chief economic value of the Cheviot is the sale of the draft ewes. Male lambs may occasionally be sold for slaughter straight from the hill, but generally they are fattened on the lower land, giving top-quality mutton.

North Country Cheviot

In 1792 Sir John Sinclair of Ulbanter, Caithness, purchased 500 Cheviot ewes from the Cheviot Hills and took them north to his farm. The sheep did well, and since then the breed has become very popular in Northern Scotland. The North Country Cheviot is a stronger sheep than the Cheviot from the Cheviot Hills; the head is white, the nose being straight to slightly Roman; the breed is usually hornless but occasionally a ram is found to be horned. The ewes make excellent mothers, milk well, and produce lambs capable of reaching 15·5 kg–18 kg carcass by August- September straight from the hill.

Welsh Mountain

This is one of the smallest mountain breeds, white-faced, and very active and hardy. The rams have horns, the ewes being hornless. The body is small, mature sheep weighing often only 36 kg–45 kg liveweight. The wool is similar to Cheviot quality (36s–50s) and suitable for best-quality tweeds, Welsh wool blankets, etc. Fleeces weigh around 0·9 kg–1·3 kg.

Recently the draft ewes have been crossed with the Border Leicester ram to produce the Welsh Half-bred ewe. Mutton quality is excellent.

Swaledale

Very similar in appearance to the Scots Blackface, but slightly larger in size. The upper portion of the face is dark, and the muzzle white or grey (mealy).

The Swaledale is rapidly gaining favour in Scotland as well as being popular in Northern England. The wool is only suitable for carpets and coarse tweeds, a characteristic of the fleece being its tight undercoat. Average fleece weight is about

1·3 kg–1·8 kg. Crossed with a Wensleydale or Teeswater Ram, the breed produces the well-known Masham ewes.

Longwool breeds

Border Leicester

This breed is found in upland areas of the North Country and parts of Scotland. The sheep are easily recognised: the head is covered with white hairs; the ears are long and lie forward; there are no horns; the nose is Roman; the legs are clean below the hock; the back is broad and long; and the sheep stands high off the ground.

The breed is famous particularly for its rams for crossing with Cheviot, Scots Blackface, and Welsh Mountain ewes. Bred pure, and under favourable conditions, this breed is capable of very high lambing percentages, many flocks averaging 180 per cent.

Wensleydale

This breed has a distinctive appearance, the head being of a bluish colour, the fleece long in the staple and finishing in knots or curls. An average clip is around 4 kg–4·5 kg. Rams are sold for crossing with Swaledale ewes, but there is evidence to suggest that the Wensleydale is losing popularity to the Teeswater for sireing the Masham ewes.

Teeswater

The Teeswater originated in the Teesdale district of County Durham, where they are still very popular today. The rams are in great demand for crossing with the Swaledale, Rough Fell and Dalesbred to produce the Teeswater Half-bred or Masham.

The breed is similar in size and character to the Wensleydale, but the head is white or grey in colour. Fleece weights of around 5·4 kg–6·75 kg are common.

Kent or Romney Marsh

The Kent breed which is found widely in South-East England is probably better known and more widely used in New

Zealand, than in this country. It is a big sheep, white-faced, and carries a good fleece, often clipping 3·6 kg–4·5 kg.

The breed is often stocked at up to twenty ewes and lambs per hectare on the Romney Marsh, and it is claimed by local farmers that the breed is resistant to internal parasites. As a mutton breed, the Kent is supreme among longwools, although the joints tend to be rather large for modern requirements.

Today many of the Kent flocks are crossed with a Southdown ram to produce the Kent Half-bred and some are crossed with the North Country Cheviot to produce the Romney Half-bred.

South Devon

The South Devon is a localised breed which is rarely found outside its native Devon and parts of Cornwall. The face is clean and white, whilst the head is covered with curly wool; the ears often have black spots.

The wool is long, dense and curly and may weigh up to 7·2 kg–7·6 kg per fleece.

Lincoln Longwool

This breed is mentioned more for interest than economic importance, since the breed is the largest of the British breeds and mature rams may weigh in excess of 152 kg. Commercial hoggets are capable of reaching carcass weights of 40 kg at twelve months of age.

The wool is long in the staple, reaching 430 mm–480 mm, and is very dense and rather coarse; fleeces weigh around 4·5 kg–5·4 kg.

Needless to say, this breed produces joints far too big for present-day consumer demand.

The Leicester

No list would be complete without mention of the Leicester sheep, though it is, unfortunately, more of academic interest than of commercial importance.

Its ancestors being the native unimproved sheep of Leicestershire, the Leicester breed was evolved by Robert Bakewell (1725–1795) of Dishley Grange, Leicestershire. By careful selection and vigorous culling, Bakewell produced a breed capable of early maturity and producing good-quality mutton. The Leicester was used to found the Border Leicester breed, as well as to improve many other breeds.

The present-day Leicester is of medium size, white-faced, the forehead being covered with wool. The fleece is fairly dense, of good length and curly, average fleece about 4·5 kg. The breed has been widely exported.

Shortwool Down

Southdown

If one is looking for the ideal butcher's lamb, short, thick and blocky in appearance, carrying a deep flesh all over, with fine textured muscle and an adequate but not excess covering of fat, one need look no further than the Southdown breed. It was one of the first of the Down breeds to be improved by John Ellman of Glynde, Sussex, towards the end of the eighteenth century. The breed has been exported to many countries and used extensively in New Zealand and Australia.

The body is small, but blocky; the head is distinguished by its mousey grey face, and short rounded ears. The fleece is of very high quality and second only to the Merino; the staple length is 50 mm–75 mm, with an excellent crimp, average fleece weight 2 kg.

Suffolk Down

Undoubtedly, the Suffolk ram is the most popular crossing sire of all breeds today. It is striking in appearance, with jet-black head, no wool on the poll, aquiline face and dropping ears. The breed originated from crossing the black-face Norfolk Horned breed with the Southdown. Suffolk rams sire lambs capable of rapid growth, show good fleshing and are light in the bone.

Provided the dam has sufficient milk, most Suffolk × lambs reach slaughter weight in 12–16 weeks. The breed does equally well when fed on cheap arable by-products, as in the Eastern counties, or kept on rich grazing pastures in the Midlands and West Country. Wool weights are around 2·7 kg–3 kg.

Dorset Down

This breed provides mainly crossing rams for early fat lamb production. The breed carried both Hampshire and Southdown blood in its early improvement. Like the Hampshire Down, the breed is suitable for folding on arable land, especially the chalk downlands of the South-West. A characteristic of the breed is its ability to lamb down in November–December.

Oxford Down

The Oxford is the largest of the Down breeds, and the rams have a majestic appearance. The face and legs are of a brownish colour. The breed is hardy and able to stand up to cold, wet weather. Today, like most of our large breeds, they are becoming less favoured by butcher and housewife, which is a great pity when one recalls how many Oxford × lambs were produced in times of meat shortage during the Second World War.

Hampshire Down

The Hampshire originated from crossing the Southdown with the old dark-faced Berkshire Knot × Wiltshire. The breed is of medium size, and capable of very rapid growth. The face is dark brown, and there are dark patches on the knees. Wool quality is similar to Dorset Down. Hampshire Down × lambs are very suitable for folding over root crops.

Shropshire

Originating in Shropshire, the breed is one of the oldest Down breeds. In size, the Shropshire is larger than its Herefordshire neighbour the Ryeland, but smaller than the Suffolk and the Oxford. The breed was used to found the Clun Forest, and has today lost favour to this relatively new breed.

The breed produces good-quality wool, the average weight is about 2·25 kg–2·7 kg.

Shortwools—not Down breeds

Ryeland

The Ryeland is one of the oldest breeds of sheep found in England. It is claimed to have been bred from native sheep kept by the monks at Leominster Priory, although other sources suggest that the breed originated on the Rye lands of South Herefordshire.

In appearance and quality the Ryeland is similar to, although rather larger than, the Southdown. The face is a dull white colour with dark nose and nostrils.

The conformation is blocky, and the breed tends to get too fat if kept on the best pastures.

Perhaps the Ryeland is most famous for its quality of wool, often found to be 56s–60s. It is ideally suited for fine hosiery, and knitting yarns; the average fleece weighs 2·7 kg.

The breed thrives on poor land with a minimum of hand feeding. The rams are becoming increasingly popular for early fat lamb production. Like the Romney Marsh sheep the Ryeland appears to be almost immune to foot rot.

Wiltshire Horn Western

The Wiltshire breed is different from all other breeds in that it carries little or no wool. Despite this, the breed is reasonably hardy and quite often found in North Wales and the Cotswolds.

The rams are popular for crossing with lowland ewes for early fat lamb production.

Dorset Horn

The Dorset Horn is a breed that is rapidly gaining in popularity owing to its ability to produce two crops of lambs a year. The ewes will take the ram at any time of the year—generally in April–May, and again in October–November. In practice, ewes usually drop three crops of lambs in two years.

Much interest has been shown recently in the polled Dorset Horn, a similar type of ewe with, of course, no horns. Lambing percentage 225 per cent with twice-yearly lambing.

Clun Forest

The Clun Forest is probably the most popular breeding ewe to be found in the English lowland. Today there are over 1000 flocks of registered ewes, with, it is claimed, a flock in every English county. The Clun has a clean face, slate brown in colour, with a slight covering of wool over the top of the head. The ears are upright, and great emphasis is paid by the breeders to the position and size of the ear. Breeders claim that the whole character of the Clun is denoted in the head. Wool quality is good, average fleece weighing 2·25 kg–2·7 kg. Flocks are generally crossed with a Ryeland or Suffolk ram for early fat lamb production, or a Hampshire or Oxford to produce store lambs for fattening on root crops.

The Kerry Hill

The Kerry Hill breed originated in the Kerry Hills of Montgomeryshire. The breed is descended from improved Welsh ewes crossed with dark-faced rams; thus the breed is particularly hardy and able to thrive under adverse hill conditions or improve rapidly in more favourable lowland conditions.

The face is white with black butterfly markings over the nostrils, and black markings are carried on the legs. The wool is quite dense, but inclined to show some kemp over the britch. Average fleece is 2·25 kg–2·7 kg.

The ewes are never docked, as it is claimed that the tail gives protection to the udder. This may, however, prove a serious disadvantage under lowland conditions, where there is a greater risk from blow-fly attacking dung- and urine-stained tails (see page 139).

The Kerry ram is widely used on Welsh ewes to produce speckled-face sheep popular in Mid-Wales. Wool 56s–58s.

Hill Radnor

The breed originated in the Black Mountains which run through the counties of Radnorshire, Herefordshire and Monmouthshire. The Radnor ewe has been popular in these and adjoining counties.

The ewes are larger than the Welsh Mountain, brown-faced, slightly Roman-nosed and hornless, whilst the rams carry horns.

The wool is similar in quality to the best Welsh, but produce a heavier fleece, approximately 1·8 kg. Crossed with a Ryeland ram, the ewes produce superb-quality fat lambs when kept under lowland conditions. Store lambs may be produced by crossing with a Clun or Shropshire sire. Lambing average under lowland conditions around 140 per cent.

Beulah Speckled Face

Although the Speckled Face has been popular in Mid-Wales for many years, the Beulah has only recently started a pedigree flock book society. Local farmers around Beulah in Breconshire have bred the Speckled Face sheep pure, and to a certain type that is popular in Mid-Wales.

The breed was developed from Welsh and Kerry sheep. We can look forward with interest to the development of this 'new' breed.

Hybrid and Half-bred sheep

Only pure-bred sheep have so far been considered, though the majority of commercial flocks are of cross-bred ewes, put to a pure-bred ram.

One may find at a local sheep auction many breeds represented in pens of ewes that are sold as either light-faced or black-faced sheep.

There are, however, a few well-known hybrid and half-bred breeds which are described below.

Colbred

The Colbred, named after its originator Mr. Oscar Colburn, was evolved from the fusion of three English breeds—the Clun, Dorset Horn, Border Leicester, and the Friesland breed, which was imported from Holland, because of its high prolificacy and milking ability.

Colbreds are white-faced, with clean, white legs. It is claimed that the breed is very easy to shepherd, prolific and deep milkers.

Scots Half-bred

The Scots Half-bred is produced by crossing a Border Leicester (longwool) ram with a draft Cheviot (mountain) ewe after she has finished her useful life on the hill. These ewes are sold to the lowlands for fat lamb production.

The distinguishing features are a white face, Roman nose, and long ears which are carried slightly forward. The legs are clean below the hocks.

Lambing percentage 150–180 per cent, and wool weight is 2·7 kg–3·1 kg.

Greyface

The Greyface is a very well-known half-bred ewe, the progeny of the Border Leicester ram, crossed with draft Scots Blackface ewes. In terms of thriftiness, hardiness and prolificacy they are very similar to Scots Half-bred.

Masham

The Masham ewe is bred by crossing Swaledale, Roughfell or Dalesbred ewes with either Teeswater or Wensleydale rams. The Masham is claimed to be hardier than either the Scots or Welsh Half-bred and is able to live on poorer land. They are variable in appearance having either black, almost white, or speckled faces. Their wool also varies considerably in quality, fleeces weighing about 2·25 kg–2·7 kg. On the lowland farm they are highly prolific and crops of up to 200 per cent have been recorded.

Masham ewes have won the National Lambing Competition for six years in succession.

Welsh Half-bred

Recently we have seen the introduction of the Welsh Half-bred, being the Border Leicester cross Welsh mountain ewe.

These ewes are smaller than the Scots Half-bred, and may be cheaper to buy. They have become extremely popular in the Midlands, especially on the Cotswolds. It is claimed by the Welsh Half-breed Association that the ewes, which are not as big as many similar breeds, are easier to handle, demand less food and are very easy to shepherd. Lambing percentage is about 150 per cent.

Mule

The Mule ewe has gained enormous popularity in the past decade. It is bred by crossing a blue-faced Leicester on hill ewes—mainly Swaledale, although sometimes Roughfell and Herdwicks are used. The resulting progeny have a brown face and slightly Roman nose. The ewes are exceptionally hardy and prolific and frequently the ewe lambs are bred from as hoggs at one-year-old.

Suffolk Cross

Numerically the most important cross-bred ewe in the lowlands is the Suffolk Cross. This is due to the great popularity of the Suffolk ram as a sire of butchers' lambs. However, there is always a ready market for a Suffolk Cross ewe lamb. Also, commercial fat lamb producers often retain a number of their ewe lambs as breed replacements.

The Suffolk Cross varies in size from the very big ewes like the Suffolk Cross Scots half-bred to the Suffolk Cross Welsh Mountain—a small hardy sheep.

When further crossed to a 'Down' ram the resulting lambs make ideal butcher's lambs.

European breeds

Île de France

The Île de France breed society was established in France 1922. The breed was formed by introducing English Leicester blood into the native Rambouillet; some Merino blood was used, also. Today, the Île de France is used as a meat sire to produce cross-bred lambs with very lean carcasses.

The breed is of medium size and has a white face with a characteristic 'pink' nose. The wool is of good quality and the average fleece weights around 4 kg:

Friesland

The Friesland breed is native to the East Friesland district of Holland. It is a large-framed breed and noted for its milking ability and proficiency.

The breed was introduced into the United Kingdom as a 'parent breed' for producing hybrid and half-bred ewes. The breed has been used most successfully in the formation of Colbred; and is also used to produce 'milky and prolific' half-breds by crossing with Romney and other longwool breeds.

Texel

There are two distinct strains of Texel sheep—namely, the Dutch strain and the French. The original Dutch Texel is both prolific and milks well, while the French strain is renowed for its meat-producing characteristics.

The French Texel was introduced to the United Kingdom for its carcass quality, and already it is producing lambs with a high lean meat content. The breed is also in Ireland where, according to trial work, excellent carcasses have been produced.

The British Texel Flock Book Society publish annually a flock book which not only includes the pedigree records but also the growth rate and classification of the physical attributes of the pedigree Texel sheep. The breed is now very well established in Scotland and its popularity is spreading throughout the U.K. The breed is characterised by its strong broad white head and clean white legs. Texel ewes are prolific, good mothers and produce an abundance of milk. Wool weights are around 4 kg.

Finnish Landrace

The Finnish Landrace is a small breed noted for its exceptional proficiency. It has a white face, pinky nose and only weighs around 45 kg to 50 kg. Although its carcass traits are poor, it is often used as a constituent of hybrid sheep to improve the lambing percentage. As a pure breed lambing averages of up to 300 per cent have been achieved.

Major crosses	*Died before lambing*	*Empty*	*Lambed*	*Total lambs born*	*Live lambs born*	*Reared*	*Live lambs born per ewe lambed*
Clun crosses	2	6	92	154	143	136	1·55
	1	4	95	176	158	145	1·66
Dorset Horn crosses	2	6	92	147	139	132	1·51
	1	4	95	168	150	140	1·58
Greyface	2	5	93	174	163	153	1·75
	1	3	96	198	179	168	1·86
Kerry crosses	2	6	92	151	142	135	1·54
	1	4	95	168	150	140	1·58
Masham	2	6	92	172	163	152	1·77
	1	4	95	196	178	167	1·87
Mule	2	5	93	176	164	154	1·76
	1	3	96	201	181	168	1·89
Romney Half-bred	2	5	93	137	129	123	1·39
	1	3	96	156	139	130	1·45
Scots Half-bred	2	5	92	171	159	150	1·73
	1	3	95	195	179	165	1·88
Welsh Half-bred	2	4	94	152	141	135	1·50
	1	3	97	173	158	147	1·63
Suffolk Crosses							
Clun	2	5	93	146	140	133	1·51
	1	3	95	166	151	141	1·59
Masham	2	5	93	173	160	152	1·72
	1	3	96	197	175	162	1·82
Mule	2	5	93	169	159	151	1·71
	1	3	96	193	176	161	1·83
Romney Half-bred	3	5	92	144	137	130	1·49
	2	3	95	164	149	138	1·57
Scots Half-bred	2	7	91	163	151	143	1·66
	1	4	94	186	167	155	1·78
Welsh Half-bred	2	5	93	152	144	137	1·55
	1	3	96	173	156	145	1·63

Source: MLC

Conclusions

At first sight it would appear to the newcomer to sheep husbandry that over forty breeds, plus the crosses, are far too many for such a small country as the British Isles. It should be remembered, however, that soil and climate vary considerably, and that certain breeds do better on their native soil whilst other breeds thrive best when moved to different ground. In choosing a flock one should:

1 Consider if the flock is intended for early fat lamb production, store lambs, or producing future breeding stock.

2 Study the farm on which you intend keeping your flock—hill, upland, lowland, grass or arable.

3 Select a breed according to your own personal preference where possible. Obviously you cannot keep Southdowns on a Scottish hill farm, but an enthusiast for a breed that may not be in fashion at present will almost certainly be successful by sticking to the breed that he likes.

4 Start by obtaining the best possible stock that you can afford. Nothing is more disappointing than breeding from poor-quality stock.

5 Always buy sheep off poorer land than your own. Moving sheep on to better land is one of the secrets of successful shepherding.

Six Selection of Breeding Stock for Fat Lamb and Mutton Production

There is a well-known saying amongst shepherds that 'when buying sheep, remember that one day the sheep you buy, or its progeny, will end up on the butcher's block'. There is much truth in this, and one should always bear in mind that the main function of sheep is meat production. When we buy our Down ram for crossing we must choose a ram with extreme mutton characteristics, and when we buy a ewe for breeding we should select those likely to be prolific, good mothers and capable of rearing 'blocky' butchers' lambs.

Ram selection

The final choice of breed will be made by the farmer, and his personal preference and past experience must be recognised. However, as a guide for the beginnner it is safe to generalise on the rams best suited for the three main lowland systems. These are:

1 Early fat lamb
2 Store lambs
3 Breeding ewes—self-contained flock

Fat lambs

There is quite a wide choice of shortwool breeds, all suitable for sireing good-quality commercial butchers' lambs. The Suffolk, Southdown, Ryeland and Dorset Down are very good examples. The lambs sired by these great breeds will be of high quality, blocky in appearance, quick to mature, and yield carcasses with a high proportion of valuable joints.

Store lambs

There is a marked preference by sheep buyers for dark-faced store lambs. Farmers believe that they have a great capacity for heavy feeding and will stand up to the stress of folding on root crops better than light-faced lambs. The Hampshire ram is widely used in flocks that aim to produce strong-boned, stocky store lambs for the autumn sales.

Breeding ewe lambs for replacement stock

Obviously, where a pure bred self-contained flock is kept, you would use a ram of the same breed as the ewes. The main consideration when selecting your ram is to look for breed characteristics, and if possible to buy a ram that was born a twin. Although twinning is not highly inherited, it can be improved by using only 'twin-born' rams over several generations.

Buying a ram

A visit to the local ram show and sale will reveal many breeds that are popular in your area. If you study each breed carefully you will soon recognise strains within the breed; the strain of sheep is as important and sometimes as variable as the difference between individual breeds. You will observe within the breed strains of 'élite' or 'breeders' sheep: these are the short-legged, stocky, compact rams that exhibit the breed's chief characteristics. On the other hand, you will also find large-framed, somewhat coarser animals that show a capacity for good growth but lack the quality of the breeders' ram. These are called crossing rams and are well suited to fat lamb production.

Inspection

Pick out several rams that are well grown, masculine, and carry their heads high. Have the rams paraded in order to see how they walk; avoid rams with sickle hocks, or those that drag their feet along the floor. A ram must have sound feet and a good walking action if he is to be able to cover a large flock of ewes. Remember that a ram is bought to get ewes in lamb and must therefore be fit, active and fertile.

Handling

It is only by handling a sheep that you are able to judge its true merit. Most rams are skilfully trimmed by shepherds for a show and sale and to the unwary they may be deceptive in appearance. By handling a ram you feel his frame and fleshing qualities. It is best to start with the head and to work systematically back to the dock.

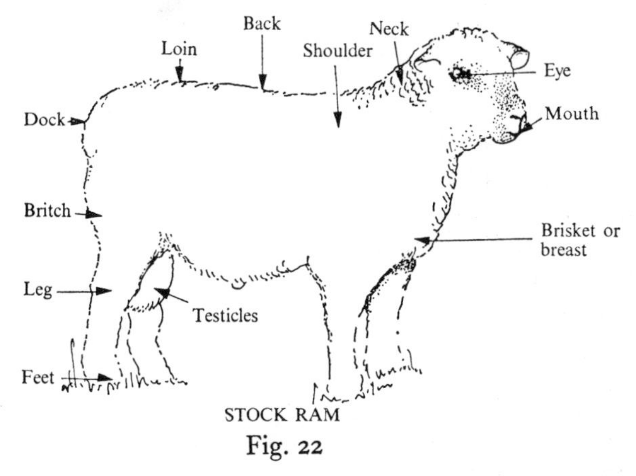

Fig. 22

The head should be masculine and held high; look for a bold bright eye with good colour; a watery eye, or one with pale membranes, indicates poor health. Open the mouth to examine the teeth and jaw. If a ram has a faulty mouth, with either an overshot or undershot jaw, or his teeth are crossed (pig-mouthed) or otherwise faulty, then leave the ram at once and find another sheep with a sound mouth. Next, move on to the neck and shoulder, and grasp the neck in between your outstretched hand; this should feel firm and be relatively short and strong. Handle the shoulder by thrusting the fingers through the wool, feel for width and placing; the shoulder should continue into the ribs without being either narrow or two heavy and coarse. The loin must be wide and covered with firm flesh, since this is a very valuable part of the sheep. Now, carefully examine the hindquarters by running your fingers from the hook bones to the tail. Many rams fault at this point, and shepherds who trim their rams will leave a surplus of wool over the tail head to cover this weakness. Grasp the inside of a leg to find the amount of flesh that is carried down to the

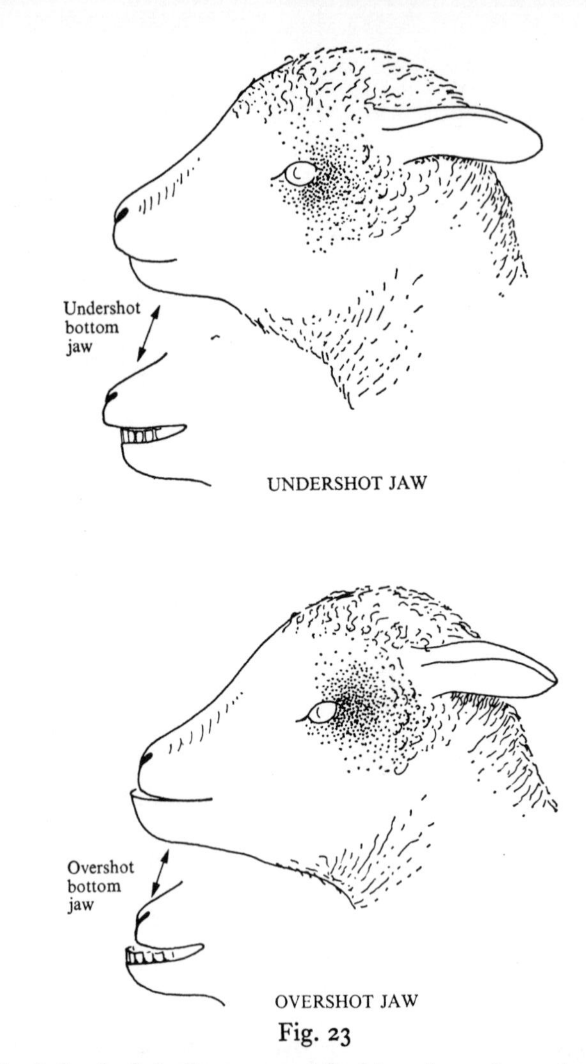

Fig. 23

hock. A fat dock indicates good fleshing throughout the sheep.

If you are satisfied with the ram's condition then move on to examine the feet and fleece. A ram is only as good as his feet will allow! Rams with well-pared even feet, free from any foot infection, must be sought. At all costs avoid rams with chronic feet ailments, or those with weak pastern joints. The importance of sound legs cannot be over-stressed.

Rams that are intended to breed future breeding ewes should be examined for wool quality since this is a way of improving

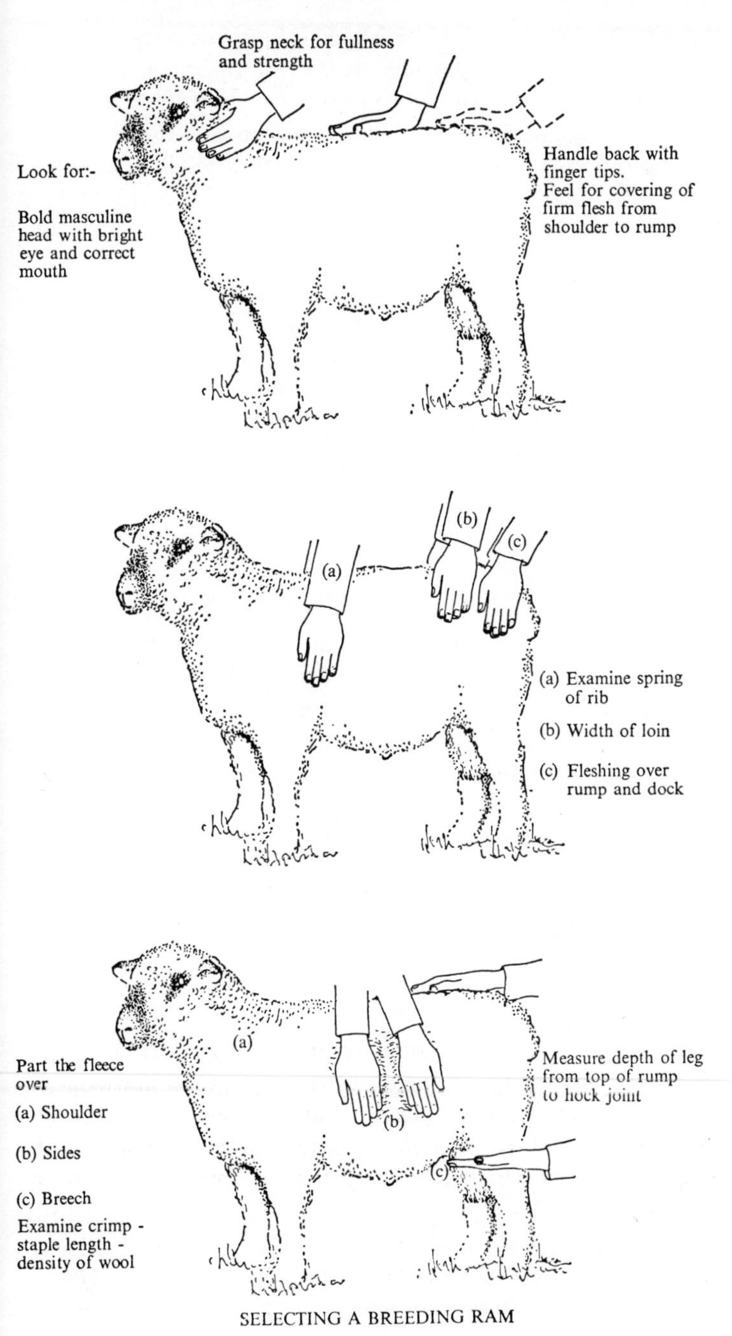

SELECTING A BREEDING RAM

Fig. 24

the wool in your flock. The best wool is found over the shoulder, and the poorest in the breech. By parting the wool at the base of the neck you will be able to see the crimp, fibre diameter and staple length (see page 154). Then move to the ribs, again parting the wool, and look for the number of folicles per square inch—this indicates the density. You can also grasp in your hand the wool on the sides as a further guide to density and strength of the fleece. Finally, move to the breech end, where you should look carefully to see if there are any kemp, grey or black fibres, which, of course, are undesirable.

It is usual for rams to be sold as warranted fertile, the buyer returning the ram if he fails to get ewes in lamb. However, it is as well to check that your intended purchase has two well-developed, descended testicles. Lastly, remember that a ram is half the flock, buy a sheep that takes your eye, and be prepared to pay a reasonable price.

Selection and purchase of breeding ewes

Before you select and purchase your breeding flock it is worth while to consider what are the qualities that you require in breeding ewes if they are to make maximum profits and fit in with your farming system.

The need for a bright eye, tight fleece, sound udder and feet are dealt with in Chapter Seven, but here we might consider some of the less obvious characteristics of the breeding ewe, such as health, size, mothering instinct, prolificacy, thriftiness, hardiness, temperament, milking ability, grazing habit, breed, character and condition.

Health

It is vitally important that livestock are healthy if they are to perform well and be profitable to the owner. Health in sheep is shown by a general alertness and an appearance of character. Healthy ewes carry their heads high, are alert and active; there is a brightness in the eye, skin and fleece that is not found in a sick sheep. Fortunately, health is so easily recognised that any ailing animals can very quickly be spotted by an experienced shepherd, the symptoms being a dull appearance, and a paleness of the skin and eyelids. Quite often a sick sheep will stray away from the flock, and may be found standing under a hedge or wall. There may be coughing, discharge from the eyes and nose, scouring and a 'pinched' appearance.

Constitution

Some ewes may be in good health, kept on good pasture and yet fail to thrive. These sheep are referred to as 'lacking constitution'. Sheep that are extremely fine-boned, narrow, with shallow bodies and stunted growth, are lacking in constitution and are unlikely to be profitable.

Select ewes with 'roomy' bodies, reasonable bone, and a sturdy healthy appearance. However, it must be said that constitution is also a reflection of a well-developed nervous and glandular system, and not entirely a matter of 'plenty of heart room'.

Body size

There is an abundance of evidence to show that the large breeds (weighing 80 kg–90 kg liveweight) produce lambs that grow more quickly and consequently, reach market weights earlier than the small breeds (36 kg–45 kg liveweight). Large breeds, such as the Border Leicester, are more prolific than small breeds like the Ryeland or Southdown. On the other hand, the larger breeds require more supplementary food in the winter and cannot be kept at such a high stocking rate in the summer.

The farmer must decide whether to keep, perhaps, ten large ewes with fifteen lambs per hectare, or fifteen small ewes with a similar number of lambs, during the grazing season.

Temperament and mothering ability

Temperament or disposition is an important characteristic. Some breeds are wild and difficult to manage; they do not respect fences or hedges and are constantly breaking out. Other breeds are so docile that they hardly lift their heads when you walk round them! Temperament is particularly important at lambing time. Good ewes will suckle their lambs readily and defend their offspring if dogs or strangers approach. Poor mothers take little interest in their lambs and sometimes leave them.

As a generalisation, we find the hill breeds to be better mothers than the Down breeds, although they may be more difficult to handle and less respectful of fences than lowland breeds.

Grazing habit

Sheep are scavengers and should be grazed on the poorer pastures, or should follow cattle. There is, however, considerable variation in the grazing habits among the breeds. Many of our upland sheep are thrifty; that is, they can make do on limited grazings, whilst the majority of the Down breeds are 'lazy feeders' and in severe weather will stand around empty troughs rather than look for food. A further characteristic of mountain breeds is that they wander away from the flock to graze. This is called open flocking, and in the course of a day the sheep walk many miles up and down the hill. In contrast, the lowland arable flocks may be folded as densely as forty sheep per acre, in small restricted enclosures.

One of the chief reasons for the popularity of such breeds as the Kerry Hill, Speckled Face and Welsh Half-breds in the lowlands is due to their winter foraging habit, in that some ewes would rather move snow with their feet in search of grass than accept hay from a rack.

Conformation and milking ability

There seems to be a connection between the conformation, fleshing qualities, ability to fatten and milking ability. The same is true of cattle where we have beef breeds that fatten quickly, yet the cows produce very little milk compared with dairy cattle such as the Jersey and Ayrshire.

Thus our extreme mutton breeds like the Southdown and Ryeland do not appear to milk as well as the more angular breeds like the Scots Half-bred and Masham. It is only fair to say 'do not appear' since this is an observation based on limited evidence, and very little experimental work has been done on milk recording in sheep.

Body condition scoring of ewes

The body condition of ewes at tupping has an important effect on the number of lambs born per ewe and the proportion of barren ewes in the flock. Condition scoring provides a convenient method of assessing this important factor, it is based on standard handling technique, and the allocation of a score in the range from 1 (very lean) to 5 (very fat).

MLC studies have shown that lowland ewes should have a condition score of about 3½ at tupping to achieve the best lambing results, but the majority of flocks contain ewes with scores ranging from 1½ to 4. Most flocks contain a significant group of ewes which would benefit from being improved in condition of mating.

The degree of condition is assessed by handling over and around the backbone in the loin area immediately behind the last rib and above the kidney, using the fingers along the top and sides of the backbone. See Fig. 25. Training in the method is available through MLC staff and other advisory services. First the degree of sharpness or roundness of the lumbar vertebrae (the bony points rising upwards from the back) is assessed: secondly the prominence and degree of cover of the transverse processes of the vertebrae (the bones coming out from each side of the backbone) is assessed. Thirdly the extent of the muscular and fatty tissue underneath the transverse processes is judged by the ease with which the fingers pass under the ends of the bones. Finally, the fullness of the eye muscle and its degree of fat cover is judged by pressing the fingers into the angle between the spinous and transverse processes.

The scoring system

Everyone working with sheep has their own personal method of assessing condition, but a problem arises when it is necessary to describe the condition of one animal, or the general condition of a flock as a whole. It is difficult to convey this information without using equivocal phrases such as 'poor', 'fairly good' or 'moderate store'. What is considered to be 'forward store' by someone accustomed to working with lowland sheep, for example, might be termed 'very good' by another more familiar with hill stock.

The system of body condition scoring has been developed as a method of overcoming these difficulties. It involves giving a score on a standardised basis, on a scale ranging from 0 to 5,

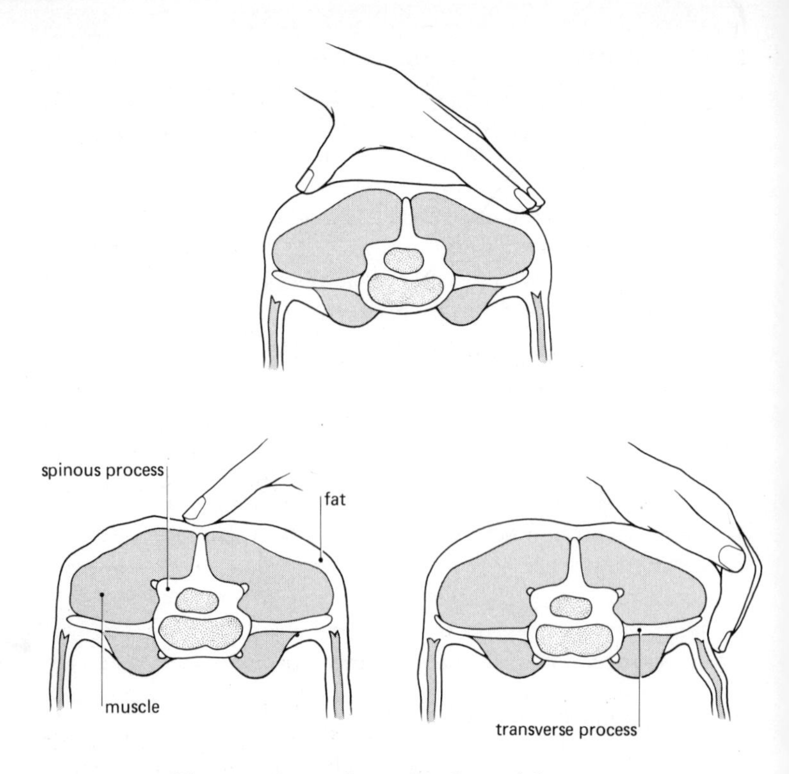

Fig. 25 **Assessing muscle and fat cover**

Ensure that the sheep is standing level and square on all four feet and is not pushing forward. Its back muscles must be relaxed.

1 Use your fingers and thumb over the loin of the sheep
2 Feel for the tips of the transverse processes: Do the bones feel sharp or smoothly rounded? How far will the tips of your fingers go under the transverse processes?
3 Feel the fullness of muscle and fat cover on either side of the spinous processes and between the spinous process and the transverse process

according to the degree of condition, assessed by handling in the prescribed manner. The system was developed from an Australian method by the Hill Farming Research Organisation, based on a six-point scale from 0 to 5. In practice points 0 and 5 are rarely used and most ewes score between $1\frac{1}{2}$ and $4\frac{1}{2}$. Scoring is made to $\frac{1}{2}$-point accuracy.

Condition scores

Grade 0 extremely emaciated and on the point of death. It is not possible to detect any muscular or fatty tissue between the skin and the bone.

Grade 1 the spinous processes are prominent and sharp; the transverse processes are also sharp, the fingers pass easily under the ends, and it is possible to feel between each process; the loin muscles are shallow with no fat cover.

Grade 2 the spinous processes are still prominent but smooth, and individual processes can be felt only as fine corrugations; the transverse processes are smooth and rounder; and it is possible to pass the fingers under the ends with a little pressure; the loin muscles are of moderate depth, but have little fat cover.

Grade 3 the spinous processes have only a small elevation, are smooth and rounded, and individual bones can be felt only with pressure; the transverse processes are smooth and well covered, and firm pressure is required to feel over the ends; the loin muscles are full, and have a moderate degree of fat cover.

Grade 4 the spinous processes can just be detected with pressure as a hard line, the ends of the transverse processes cannot be felt; the loin muscles are full, and have a thick covering of fat.

Grade 5 the spinous processes cannot be detected even with firm pressure; and there is a depression between the layers of fat in the position where the spinous processes would normally be felt; the transverse processes cannot be detected; the loin muscles are very full with very thick fat cover.

The main advantages of body condition scoring are the ease with which it can be learned and used and the fact that it does not require any equipment. It overcomes the problems of differences in size and skeletal shape of ewes which affect bodyweights, and it can be used in situations where bodyweights are difficult to interpret—for example, in pregnant ewes.

Sheep records

The obvious way to select sheep is to buy physically sound ewes which have records of their ancestry and performance. Ewes bred from flocks with a high lambing percentage, rapid recorded liveweight gain, and good carcass gradings are the ones to improve flock performance. The Meat and Livestock Commission operate a National Sheep Recording Scheme for both commercial and pedigree flocks. The resulting information is proving of immense value to both the individual farmer and the sheep industry.

Conclusions

In selecting your flock, look for the main economic factors that will make sheep farming profitable on your farm. Healthy, deep milking, prolific ewes mated to a well-grown Down ram, should produce really first-class fat lambs. Whilst there is much truth in the old saying that a sheep's greatest enemy is another sheep, you will find that to be profitable today you must keep sheep intensively. This calls for control over parasites and disease, improved feeding methods and, above all, a really good standard of shepherding.

Seven The Shepherd's Calendar

In this chapter we shall discuss the management throughout the year of the lowland flock that lambs in late February to early March. This is generally referred to as the Shepherd's Calendar. Basically the day-to-day routine work is similar in all flocks, whether we manage Blackface sheep on the rugged hills of Scotland or keep pedigree Suffolks in the lowlands. The shepherd's year begins by making up the flock.

Making up the flock

Making up the flock is a most important task, and should be undertaken by the farmer and his shepherd during the late summer or early autumn. The job entails a careful examination of every sheep in the flock to ensure that it is suitable for breeding and will be profitable to keep. Every ewe must be caught and turned up for inspection of feet, teeth and udder. Lame sheep and those with 'broken' mouths (i.e. faulty teeth) are unlikely to be profitable, whilst ewes with faulty udders will be unable to rear their lambs and should therefore be culled from the flock.

Inspection

Udder

Having turned up the ewe, gently probe the skin of the udder to see if there are any hard lumps or swellings, which would indicate a previous history of mastitis. The skin should be soft and pliable. Examine the two teats to see that they are normal and free from injury; occasionally a ewe may be found with teats damaged or lost through careless shearing.

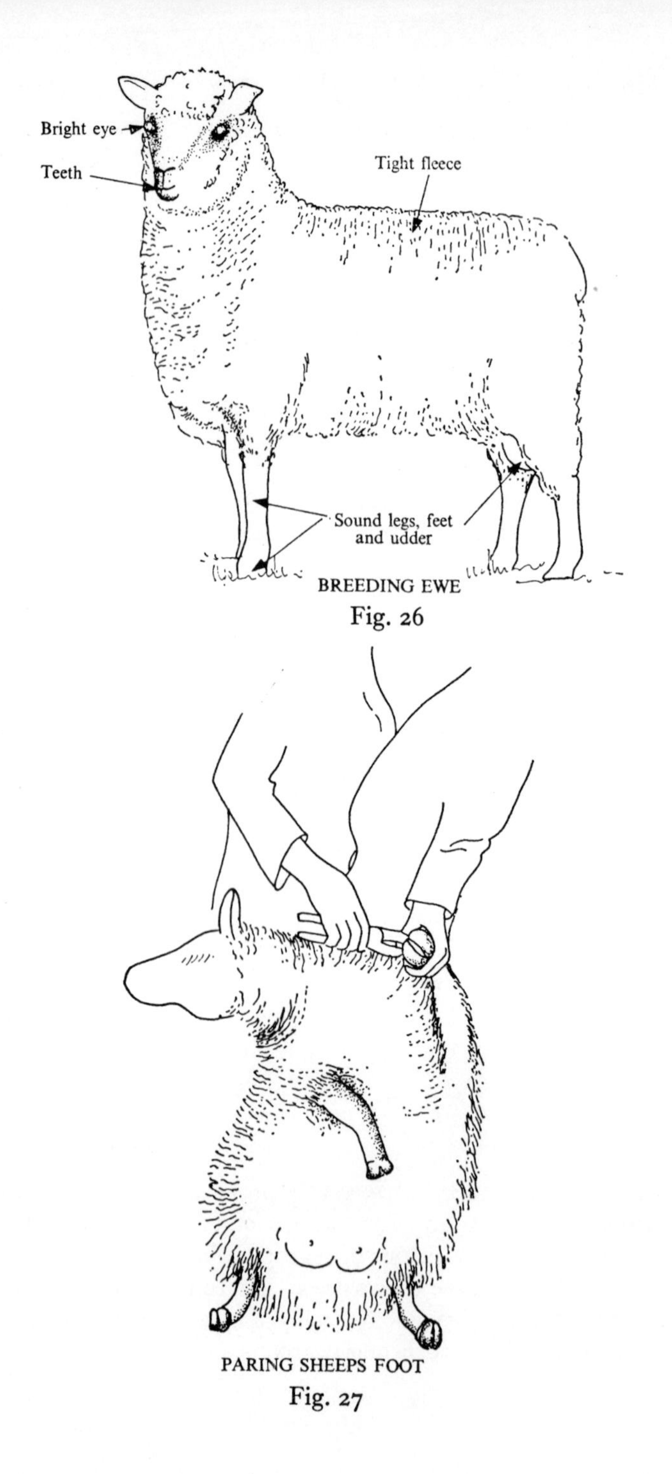

BREEDING EWE

Fig. 26

PARING SHEEPS FOOT

Fig. 27

Feet

Carefully check each foot to ensure that it is sound and firm. Overgrown nails should be trimmed off with a sharp knife or

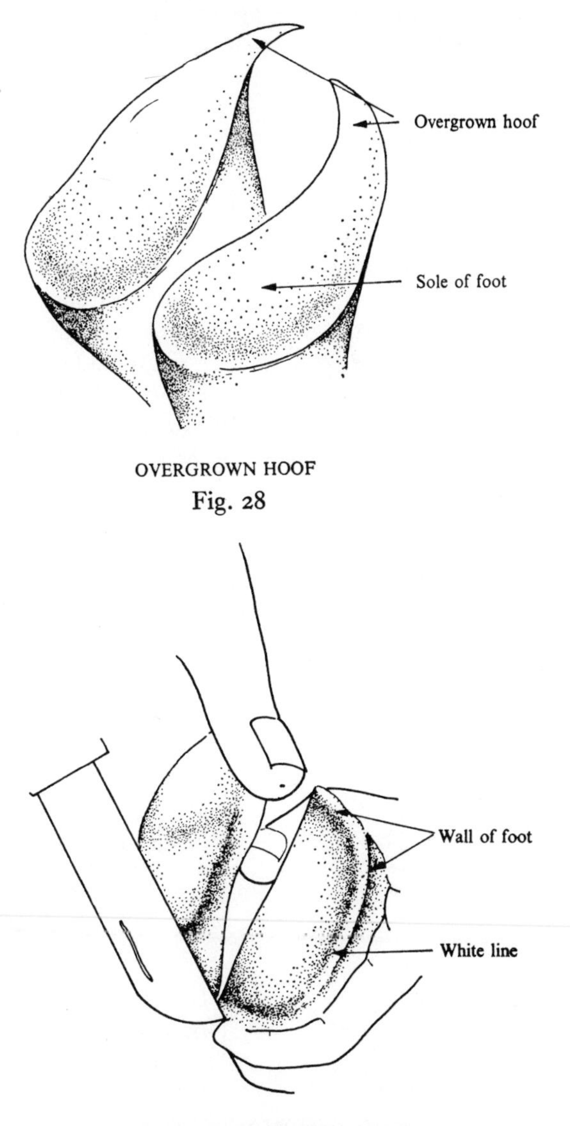

OVERGROWN HOOF

Fig. 28

FOOT PARING WALL OF FOOT UNTIL WHITE LINE APPEARS

Fig. 29

foot secateurs. If any sheep are found with foot rot they will require immediate treatment (see page 167). It is a wise policy to run the entire flock through a foot bath, containing either 6 per cent formalin or 10 per cent copper sulphate, and then turn the sheep on to a clean, rested pasture.

Teeth

Young breeding ewes are always healthier and require less food to keep than old 'broken-mouthed' sheep. The term 'broken mouthed' covers any fault of the mouth, but usually refers to sheep that have lost their incisor teeth. Since the majority of sheep keep all their permanent incisors until they are five to six years old, we associate broken mouths with advancing age. With commercial flocks it is advisable to cull ewes if a fault is found in the mouth, for obviously a sheep with imperfect teeth will have greater difficulty in obtaining food.

The approximate age of a sheep may be determined by examination of the incisor or 'broad' teeth.

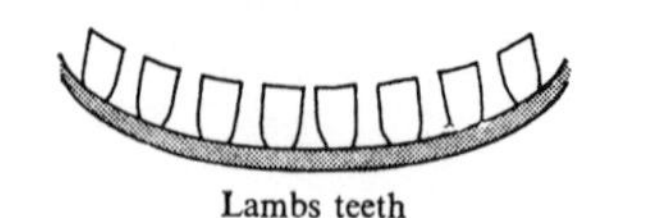

Lambs teeth

Two teeth
year old

Four teeth
2 years old

Six teeth
3 years old

Full mouth
4 years old
and more

Broken mouth
aged sheep

DENTITION OF SHEEP
Fig. 30

General health

Having satisfied yourself about feet, teeth and udders, you should next examine the general health and breeding performance of the flock. The best indication of health is a bright eye and a tight fleece. Ewes with dull, watery or sunken eyes are suspect. A loose or shedded fleece also indicates poor health or a previous illness. Quite obviously these sheep should have their ailments diagnosed and be treated medically. The fault may be traced to a previous difficult lambing or they may be carrying a heavy burden of parasitic roundworms (see p. 190). Provided that these sheep are treated, and they respond, then they may be kept in the flock. Ewes that are over-fat are always a problem, for sometimes they become erratic breeders. The aim should be to manage ewes, particularly after weaning, so that they remain fit without becoming over-fat.

Lastly, when making up the flock, remember that we have a real opportunity of bringing about an improvement in breed character and flock type. We can cull out coarse-featured animals, or those with poor-quality wool. However, always remember there is little purpose in culling a sound ewe unless you can replace her with something better. The selection of breeding ewes is dealt with more fully in Chapter Six.

Preparation of Ewes

Flushing

Prolificacy in sheep is dependent upon health, environment and breeding. Some breeds, for example the Masham, are more prolific than others. However, it is well recognised by sheep farmers that if breeding ewes are kept on good pastures for about two weeks before the rams are turned out, and they continue to be well fed during the tupping period, there will be an increase in the number of lambs born. This system is called 'flushing', and is due to the ewes being in a rising condition at the time of service. The ovaries shed more eggs, and this results in a higher conception rate. It is always easier to 'flush' ewes if they are kept in hard condition after weaning.

It is most important that the ewes continue to receive a plentiful supply of food for two or three weeks after tupping.

This will help to avoid foetal reapsorption and increase the lambing potential. Many farmers in fact feed 0·25 kg of concentrates per ewe per day during flushing and for two weeks after tupping.

The gestation period (pregnancy) for ewes is 147–51 days, or approximately five months. The rams, therefore, should be turned in with the ewes five months before you wish the lambing season to commence. The majority of lowland flocks lamb in late February–March and early April, depending upon the district and climate, although a few farmers specialise in early fat lamb production and lamb the ewes in late December and January.

Oestrous cycle

Well-grown ewe lambs become sexually mature when about six to eight months old. They come 'on heat' for 1–2 days, when they will accept the ram. The interval between heat periods is approximately sixteen days. The timing of the oestrous cycle varies with breeds: Suffolk ewes will come on heat from late July to January, whilst hill breeds like the Cheviot rarely ovulate before late September–October, and may only complete two or three cycles. The Dorset Horn breed have oestrous cycles all the year round, which allows them to produce two lamb crops a year.

Preparation of ram

It is most important that stock rams are carefully managed prior to, and during, the mating season. Rams must be fit and active if high conception rates are to be obtained. On no account should a ram be allowed to become over-fat, as this will lower his fertility. Special attention should be paid to the feet; a lame ram will have great difficulty in covering a large flock.

Raddling

Before turning a ram with the flock he should be raddled, or fitted with a sire sine harness. Raddle powder (iron oxide) may be purchased from an ironmonger in a variety of colours, and should be mixed with oil or grease, then smeared over the breast bone. When the ram serves a ewe he will leave a mark over the rump.

Alternatively, a sire sine harness fitted with a coloured crayon may be used. The crayons may be purchased in four colours. It is best to start with yellow and then change the colour every sixteen days:

Turning out	16 days	yellow
17 days	32 days	red
33 days	50 days	blue
If necessary		green

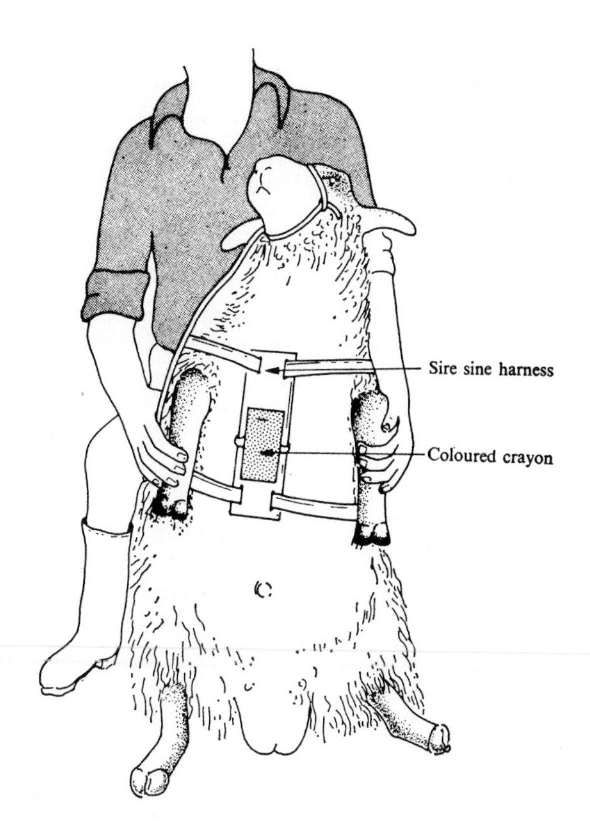

PREPARATION OF RAM

Fig. 31

By changing the colour every sixteen days the lambing date can be calculated. Also, if a ram serves a large proportion of the flock twice, it indicates his infertility, and he should be replaced immediately.

Ewes per ram

It is extremely difficult to state with accuracy how many ewes a ram is capable of serving during the mating season. Much will depend upon field conditions, breed of sheep, general health, etc. However, as a guide we may safely reckon on well-grown ram lambs serving 25–40 ewes in their first season. Mature rams should be allowed 50–60 ewes, although some rams are capable of serving more than 100 ewes.

Service

Under field conditions the ram will discharge up to ½ c.c. of semen, containing approximately 500 million sperm per service. The ewe's heat period lasts 1–2 days, and normally the ovaries shed the egg(s) towards the end of the period. Thus, ideally, the ram should serve the ewe on the second day. In practice, of course, we have no control over mating under field conditions.

The rams should be run with the ewes for three consecutive heat cycles, which means that the ram is with the ewes for seven weeks. If the rams are then removed from the flock a definite lambing period can be calculated.

Syncro mate

Controlled breeding has recently been made possible with the use of a vaginal sponge impregnated with progesterone and placed inside a ewe for a certain period before service.

In the normal oestrous cycle the follicle growth and development of the egg occur under the influence of a follicle-stimulating hormone known as F.S.H. As follicle growth accelerates, oestrogen is produced which in turn causes the symptom of heat, i.e. when the ewe will accept the ram. At this stage a leutenising hormone referred to as L.H. is produced, and this causes ovulation. After ovulation (release of egg), a corpus luteum develops and secretes progesterone which limits the follicle's activity until just prior to the next heat period.

Thus we can see that the presence or absence of progesterone determines the time at which the oestrous cycle takes place.

This is the principle behind synchronisation of sheep breeding. By administering progesterone through a vaginal sponge, we find that the oestrous cycle can be controlled.

Method

An applicator and supply of sponges are obtained from a veterinary surgeon. Rubber gloves should be worn.

An assistant should control the ewe by standing it in a race or holding against a fence or wall. Place a sponge into applicator, with loop uppermost; very gently insert applicator 100 mm–150 mm into the vagina. The sponge is then ejected by gently pushing the plunger forward, and the tube is removed, leaving the drawstring outside the vulva.

Wash the applicator and disinfect before moving to the next ewe.

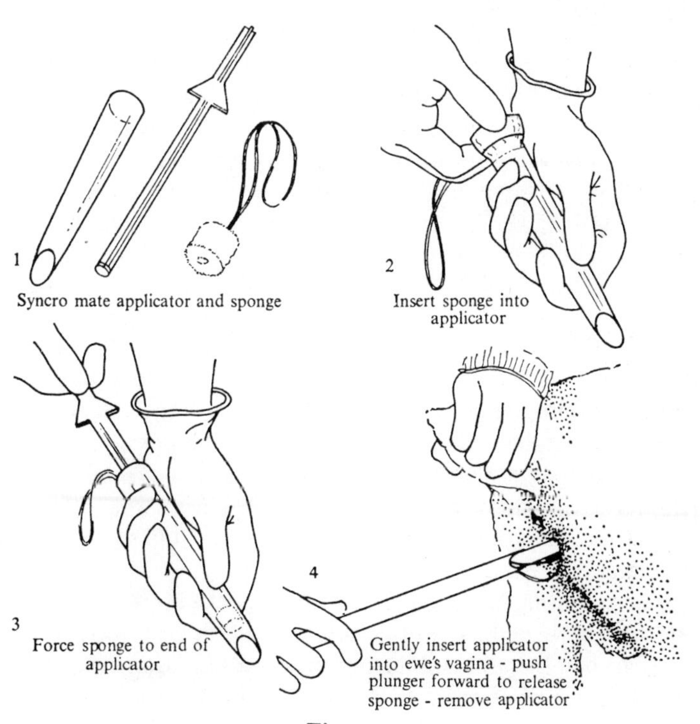

Fig. 32

The sponge is left in position for seventeen days, and then carefully removed by gently pulling the drawstring. The rams are turned with the ewes after the sponges are removed, and mating should take place during the next three or four days. If possible, a ram should be allowed ten to fifteen ewes to obtain high conception rates.

The advantage of synchronised breeding is that nearly all your lambs are born within a week, which means that fostering orphans is much easier, since you have more foster mothers available. With large flocks you can get extra help for just a week or so, and your lambs should all be roughly the same size, which gives them an even chance of obtaining food.

It is also possible to lamb ewes down 'out of season' by using progesterone sponges in conjunction with an injection of pregnant mare serum-PMS, thus getting two crops of lambs per year.

Care of the in-lamb flock

During the first three months of pregnancy the ewes should be kept in fairly hard condition, neither gaining nor losing weight. The ewes should act as scavengers, clearing up stubbles, leys, or perhaps sugar-beet tops if these are available.

Once the weather breaks, the ewes should be given some hay and moved on to the drier parts of the farm: nothing is worse for pregnant ewes than having them standing about on wet, muddy fields. In flocks where lambing takes place March–April it must be remembered that the majority of the gestation period will take place when little or no grass is available. To this end it is important to feed the ewes on the best quality hay available and to give supplementary concentrates in the latter stages.

The day-to-day management involves looking at the flock once a day to see that all is well, and checking the feet periodically to see that they are well pared and free from any infection. The flock should be moved from field to field as much as possible.

The condition of the ewes should be checked by scoring them approximately eight weeks before lambing is due, so that feed supplies can be adjusted during the steaming up period. Ewes with a body score of $2\frac{1}{2}$ or less should be separated from the main flock and given preferential treatment.

Steaming up

As the ewes become 'heavy in lamb' they will require more nutrients to feed the developing lambs. The feeding of supplementary concentrates during the last two months of pregnancy is strongly recommended. This will lead to stronger and heavier lambs at birth, the ewes will milk better and the risk of pregnancy toxaemia (see page 180) will be considerably reduced. The concentrates should be fed in troughs which will need moving daily. Allow at least 0·5 m of trough space per ewe to prevent them knocking each other when feeding.

In practice the effect of steaming up can be achieved by feeding:

8–6 weeks before lambing:	112 g concentrates + hay/head/day
6–4 weeks before lambing:	225 g concentrates + hay/head/day
4–2 weeks before lambing:	337 g concentrates + hay/head/day
2–lambing:	450 g concentrates + hay/head/day

Suitable mixtures would be:

2 parts oats
2 parts sugar beet pulp
1 part ground nut cake plus minerals

Clean, fresh water should always be freely available.

Preparation for lambing

A successful lambing season depends very much upon the preparations made by the farmer well in advance of the actual lambing. The wise farmer will keep some of his better pastures rested until lambing approaches, and where possible will have some 'early bite' for the ewes once they have lambed. The actual lambing field should, ideally, be as close to the house as possible. It should be fairly level and free from hazards; any ponds or deep ditches should be fenced off, and the field should be free-draining and provide some form of shelter from the prevailing wind. If we think of the tremendous change in environment for the lamb at birth, as it leaves the warm uterus and is thrown into the cold outside, we can appreciate the need for protection during bad weather.

Lambing shelter need not be expensive. A few straw bales and some stakes will make an ideal pen, or three sheep hurdles with some heavy hessian sacks tacked on the side and placed in a triangle will afford sufficient protection from strong winds. See Fig. 77 page 168.

Ewes which have a difficult lambing, or produce weakly lambs, should always be penned for at least twenty-four hours after lambing. This allows the ewe to recover and the shepherd to make sure that the weaker lambs are sucking.

Every shepherd should have a well-equipped 'lambing bag' in readiness for the lambing season. Some farmers provide a shepherd's hut where the shepherd stores his equipment, and where provision is made for warming milk, etc., for weakly lambs.

A suggested list of equipment is as follows:

Lambing oils, antiseptic cream, soap flakes	For lubricating hands and arms before examining a ewe internally
Nylon lambing cord	
Lambing drench	Stimulant and pain killer for ewes that have had hard difficult lambing
Soap, towel, clean bucket, approved disinfectant	For washing before and after lambing a ewe
Hypodermic syringe, quantity of 16 mm 18 swg needles, cotton wool, surgical spirits	Injecting ewes/lambs with vaccine-serum
Vaccine-serum-penicillin	Consult veterinary surgeon
Elastrator and rubber rings	Castration and tailing
Ear notcher-tagger and tags or tattoo set, marking paint	Identification of lambs
Storm lantern or torch with spare batteries	Inspecting flock at night
Feed bottle, lamb teats, milk substitute, stove	Feeding orphan lambs
Aerosol aromatic spray	Fostering orphan lambs
Notebook and pencil	Records
Dagging shears, pocket knife	Cleaning up dirty ewes and foot treatment

Lambing

A ewe that is about to lamb is restless and usually wanders off to find a sheltered place away from the flock. She lies down, but turns her head in a characteristic manner, looking up almost directly at the sky. As the labour pains increase, she makes a tremendous muscular effort to force the lamb from the uterus to the vaginal passage; this is what we understand by 'strain'. As the ewe continues to strain, she forces the 'water bag' (which until now has been used as a cushion to prevent injury to the developing lamb) into the vagina. The bag bursts and releases the fluid. The next stage is that the lamb's forefeet now appear at the vulva, and then the legs with the head resting on the knees. As soon as the head and shoulders are through the vulva opening, the main difficulty is over, and the lamb is quickly born. The navel cord snaps as the lamb drops to the ground, and the ewe will almost immediately stand up and turn round to lick her lamb.

The lamb will shake its head and quickly respond to the massaging effect of the ewe licking its skin. In a few minutes the lamb will rise to its feet, and, provided it is strong and healthy, will be suckling the ewe within a quarter of an hour or so. If the ewe has twins or triplets she will lie down again and deliver the second lamb more quickly and easily than the first-born.

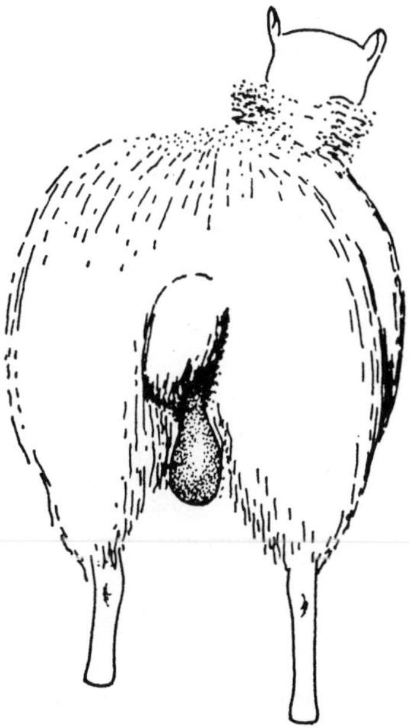

APPEARANCE OF THE 'WATER BAG'

Fig. 33

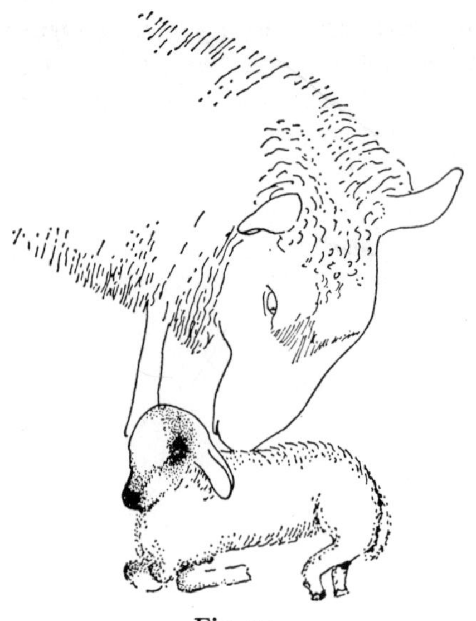

Fig. 34

A small percentage of the flock will need assistance at lambing, and the following remarks are intended only as a guide for the beginner, who, when he finds a ewe in difficulties, should always seek advice and help either from an experienced shepherd or from a veterinary surgeon. The three golden rules for difficult lambing are:

patience
hygiene
gentleness
plus common sense

Always allow ewes plenty of time before you interfere with them. At least one hour should be allowed from the time the ewe commences to strain. Young ewes in particular should be given plenty of time.

If you find the ewe does not progress, then catch her quietly. Wash your hands thoroughly, then gently insert a finger into the vagina and feel for the forefeet and nose. If these are found in the normal position, then the trouble is probably a very big lamb. You can assist the ewe by gently pulling the forelegs, first one and then the other, as the ewe strains. Once the head appears, the direction of pull should be downwards. As soon as the lamb is born, clear its mouth and nostrils of mucus, and see that it is breathing. If not, blow into the mouth, smack its ribs and work a foreleg up and down. This will usually start the lamb breathing. Place the lamb by its mother's head, then leave the pair alone for another hour or so.

Malpresentations

Again it must be stressed that a novice should under no circumstances try to correct a lamb wrongly presented unless accompanied by an experienced shepherd or veterinary surgeon.

Examining a ewe

1 Lay the ewe on her right-hand side and clip away any dirty wool from around the anus and vulva with a pair of sharp dagging shears. If an assistant is available, he can hold the ewe in position by placing a hand under the ewe's jaw.
2 Wash your arms and hands thoroughly, paying particular attention to finger nails (which should be kept short during the lambing season).
3 Work up a soapy lather on your hand and arm, or apply lambing oils or an antiseptic cream. This is to provide lubrication and sterile conditions when you examine the ewe internally.
4 Keeping your hand in a 'cupped' position, gently insert it into the vagina, enter the womb and find the lamb.
5 Diagnose the lamb's position, and change this into the normal presentation: that is, with nose resting on knees, forelegs together and stretched forward.
6 Gently draw the lamb upwards until it passes through the cervix and enters the vaginal passage. Then pull downwards until the lamb comes away. Remember only to pull when the ewe strains.
7 Ensure that the lamb is breathing and that the mouth and nostrils are free from mucus membrane.
8 Place a pessary in the womb to prevent infection, or inject 6 c.c. penicillin (see vet).

Normal presentations

Normal presentation—leave sheep alone.

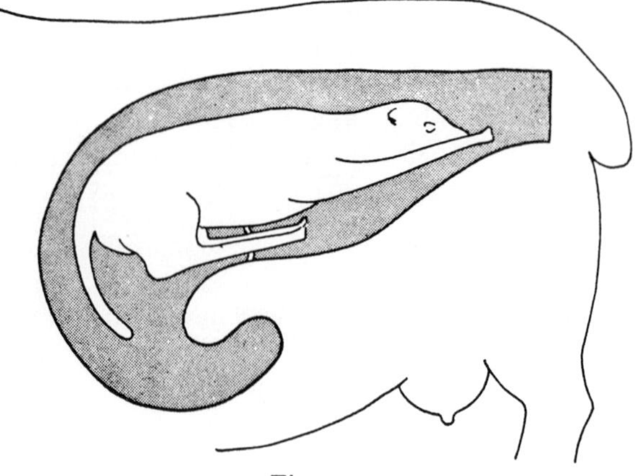

Fig. 35

Malpresentations

A few fairly common malpresentations are illustrated below, but it must be emphasised that these are by no means the only complications that may arise. A good rule at lambing is 'If in doubt, call the veterinary surgeon out'.

Leg turned back—gently push lamb into womb, draw the leg into correct position and gradually pull lamb forward, only pulling as the ewe strains.

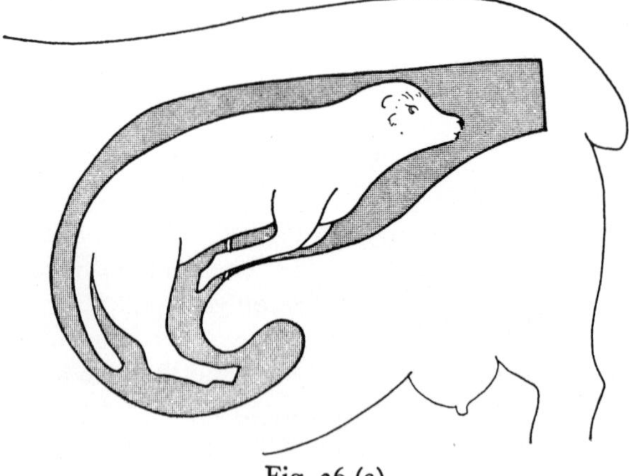

Fig. 36 (a)

Head turned back—gently move hand around head and draw into correct place. Often the head slips back, and it may be necessary to slip a sterile string round the head to keep it in the correct position.

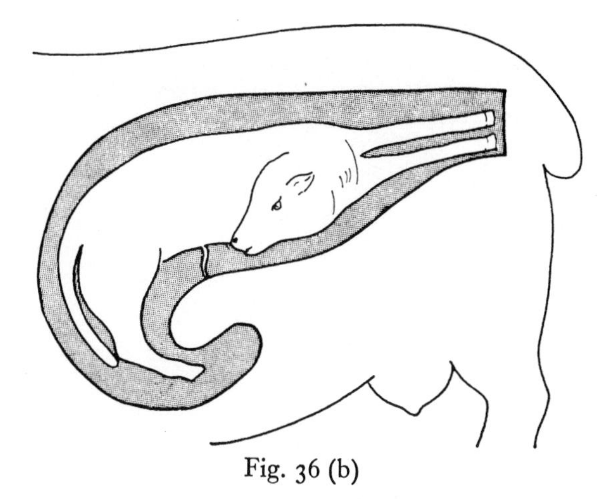

Fig. 36 (b)

A breech presentation—find the tail and then draw lamb backwards by pulling the hind legs gently upwards and then downwards as the lamb passes through the vaginal passage.

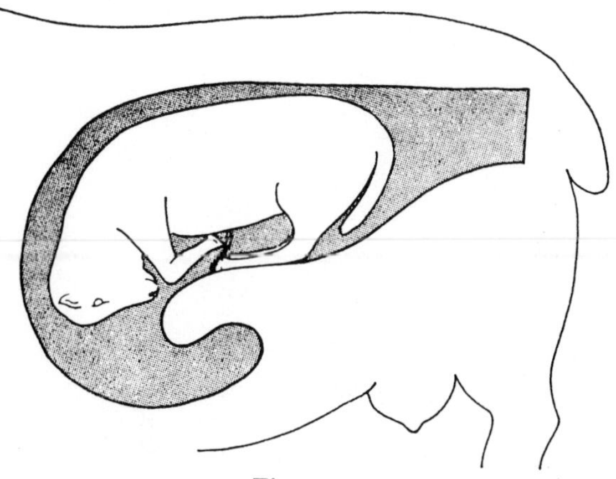

Fig. 37

Twins—both with a leg turned back—very difficult to decide. Push one lamb back, and when absolutely certain you have the correct lamb, draw gently into vagina. Second lamb will come more easily.

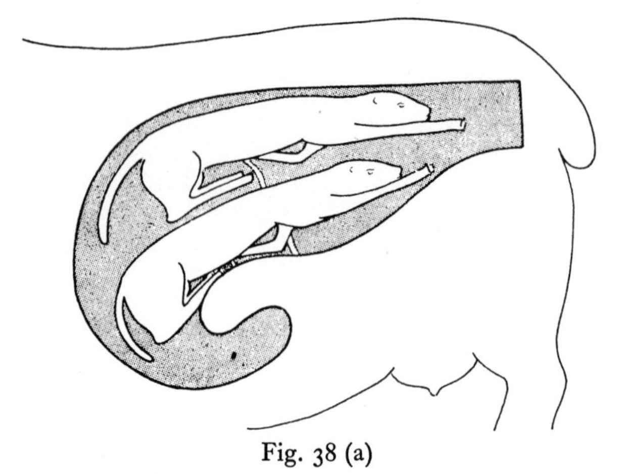

Fig. 38 (a)

Twins with breech delivery—where a breech is suspected, try to find the tail; this will make it quite certain.

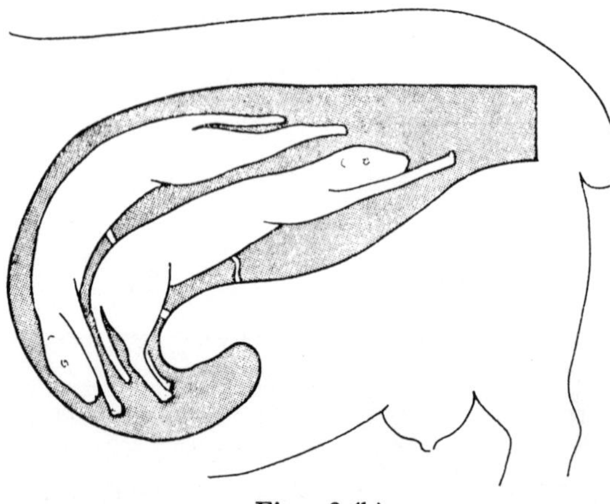

Fig. 38 (b)

The Dalton Y lambing aid may be used instead of a light cord. Place the loop over your hand, then place the loop over the lamb's head with its front legs over the Y—tighten the cord then gently draw the lamb forward by pulling the fork with your one hand, whilst the other hand guides the lamb into the vaginal canal.

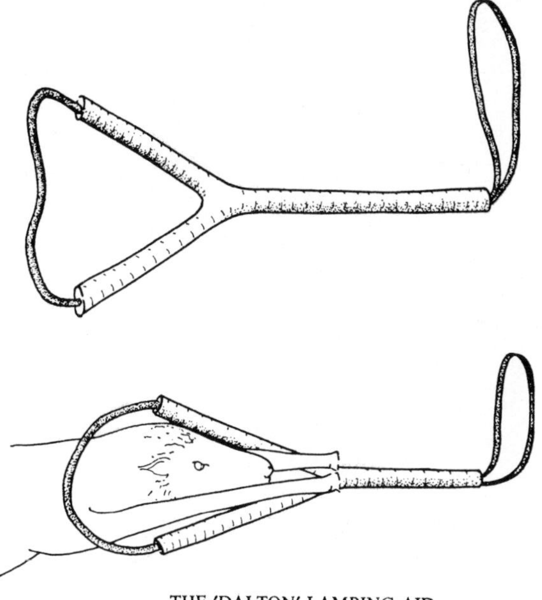

THE 'DALTON' LAMBING AID

Fig. 39 (a) and (b)

Fostering orphan lambs

The motherless lamb is one of the many problems the shepherd has to deal with at lambing time. Sometimes a ewe refuses to take to her second lamb; and there are ewes with faulty udders (who were missed when making up the flock, or later developed mastitis) who are unable to suckle their lamb, and, of course, there is the odd ewe who will die at lambing, or just after.

Where possible, the orphan lamb should be fostered by a ewe that has lost her lambs, or a ewe with a single lamb and plenty of milk. Much skill and a great deal of patience is needed by the shepherd at this time. A few tips are mentioned here as a guide.

If a ewe lambs and delivers dead or weakly lambs that die soon after birth, bring her indoors immediately. Introduce the orphan lamb as soon as possible, and rub the orphan's back

THE ORPHAN LAMB

Fig. 40

with fluid from the ewe's cleansings, if possible. You may try tying the lamb's legs loosely with string so that it will have difficulty in rising to its feet. Thus the lamb impersonates the action of a new-born lamb, and may be accepted by the ewe immediately.

A very good method which may be used in conjunction with the above is to spray the ewe's nostrils with an aromatic spray. The lamb is also sprayed, and the ewe may think that the lamb has her 'scent' and accept the orphan immediately.

The age-old method of skinning the dead lamb and tying the skin over the orphan is not recommended because the dead lamb may have been infected with lamb dysentery, or some other disease, which could be passed on.

Ewes should always be penned when making them adopt lambs, although they may be allowed out by day for a few hours to graze. It may be necessary to tie the ewe up in the first stages to allow the lamb to suckle. A very old tip is to bring a sheepdog into the pen on the second day of fostering. The ewe will defend 'her' lamb, and will usually accept it from then on. Later, the shepherd can turn the ewe with her adopted lamb out to grass, preferably in a small paddock near the house where he can keep an eye on them. He should not turn the pair with the main flock until he is quite certain that the ewe has taken to the lamb.

Fostering orphan lambs is very much an art, and is not always successful. In the event of not having a suitable ewe to act as foster mother, the lamb will have to be fed by hand.

Artificial rearing

Artificial rearing demands considerable patience and skill on the part of the shepherd. The secret appears to be feeding the lamb a little and often to avoid stomach upsets, which can quickly lead to scours (diarrhoea). A great deal of research has been carried out at the Experimental Grassland Research Institute, Hurley, into feeding lambs artifically from birth, and there is now evidence that this can be done satisfactorily with milk substitute. In general, the lambs are given four feeds per day, each feed being 0·5–0·75 l of liquid (water plus milk equivalent). Hay and concentrates should be available at all times, although the lambs are unlikely to eat much dry food until the liquid feed is reduced. Artificial rearing is rather expensive, and so far has often produced inferior results, but it is likely that considerable improvements will be made in the future.

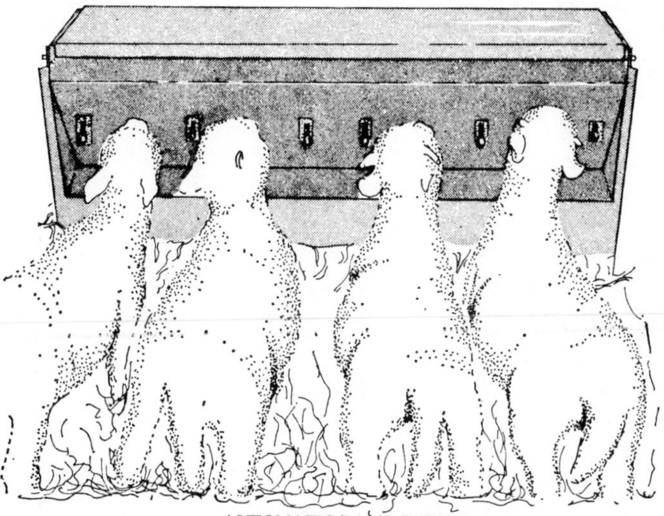

AUTOMATIC LAMB FEEDER

Fig. 41

Ewes and lambs

A ewe with her lamb is called a couple; a ewe with twin lambs form a double. If a number of small fields are available at lambing time, the couples can go in one field, and the doubles can go in another. Obviously the doubles will need extra feed, and should be given the better pastures.

Single lambs will begin to graze and nibble at concentrates when about three to four weeks old. Twin lambs will eat grass a little earlier than this, especially if the ewe is short of milk. A careful eye should be kept for ewes that are reluctant to allow their lambs to suckle, and for hungry 'pinched' lambs. The ewes should be examined for udder injury, mastitis, etc., and treated accordingly. It may be, however, that the ewe is just naturally short of milk, in which case feeding the lamb some cow's milk from a bottle will help.

Ewes that lamb late in the season (April–May) often produce too much milk for their lamb. This is caused by the spring flush of grass, and may make it necessary to strip some milk from the ewe's udder every other day for about a week. Lambs may scour a little for a day or so slightly, but they will soon settle down and be able to take the milk. This trouble is most common with ewes with single lambs.

Feeding—ewes

If ewes are well cared for during late pregnancy and are 'steamed up' on concentrates, they should lamb down with vigorous lambs, and have a plentiful supply of milk. Our aim after lambing must be to keep the ewes milking as well as possible.

Ewes with single lambs should be allowed 0·45 kg–0·9 kg of concentrates a day plus all the hay and grass they wish to consume. Ewes with twins or triplets should be allowed up to 1·4 kg of concentrates for the first two to three weeks, and then gradually cut down. A clean supply of water is vital to lactating ewes, and one or two mangolds, if available, may be fed with advantage. Mangolds contain approximately 90 per cent water, and are extremely valuable to the milking ewe.

A suitable daily ration for a ewe (47–50 kg liveweight) and twin lambs would be:

0·9 kg concentrates
3 kg–4·5 kg mangolds
ad lib. leafy hay (probably the ewe will eat about 0·9 kg)

Creep-feeding lambs is an accepted practice today. It may be achieved in two ways: for early born lambs, concentrates are fed in specially constructed lamb creeps, whilst April-born lambs may be creep-grazed ahead of the ewe flock.

To encourage lambs to eat concentrates at an early age, a fresh, appetising concentrate mixture must be used. Young lambs are particularly fond of locust beans, peas and flaked maize:

1 part flaked maize
1 part locust beans
½ part peas
2 parts rolled oats
1 part linseed cake

It is most important that the food is fed fresh, in clean troughs. Any food that is left from the previous day should be removed and given to the ewes. The ration should be fed *ad lib.* until the lambs are eating 0·45 kg per head per day.

Creep grazing Creep grazing is a system that allows the lambs to graze in front of the ewes, which means that they are able to get the young, unsoiled, nutritious grass. Creep grazing is not an easy way of shepherding. It calls for very sound judgement on the part of the shepherd. Some of his problems are how many ewes should he keep per hectare? How often should he move the flock to the next paddock? Broadly speaking, the

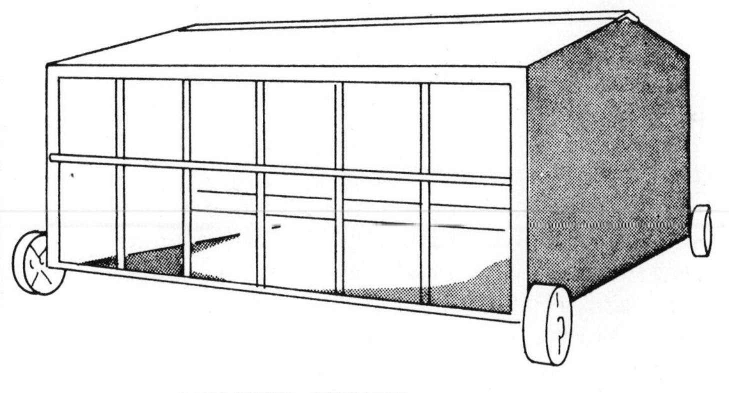

LAMB CREEP - PORTABLE
Fig. 42

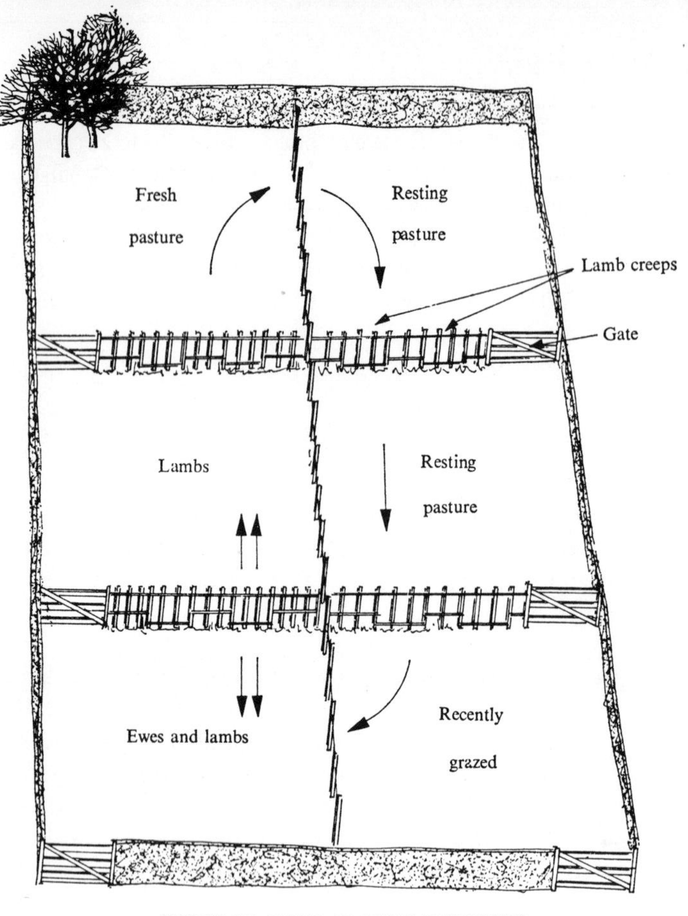

FORWARD CREEP GRAZING PADDOCKS

Fig. 43

system is as follows: A clean, healthy, well-drained pasture should be selected; this field is then divided into six equal-size paddocks, each paddock having its own creep and water supply. The field should be manured with 400 kg–500 kg of basic slag the previous autumn, plus potash if required. In the spring, the field is chain harrowed and the fences erected.

The actual date when the flock is introduced will depend upon the season, but on no account should the fields be grazed until the grass is really growing. The flock is stocked at between fifteen and twenty couples per hectare, depending upon the breed of ewe and the quality of pasture.

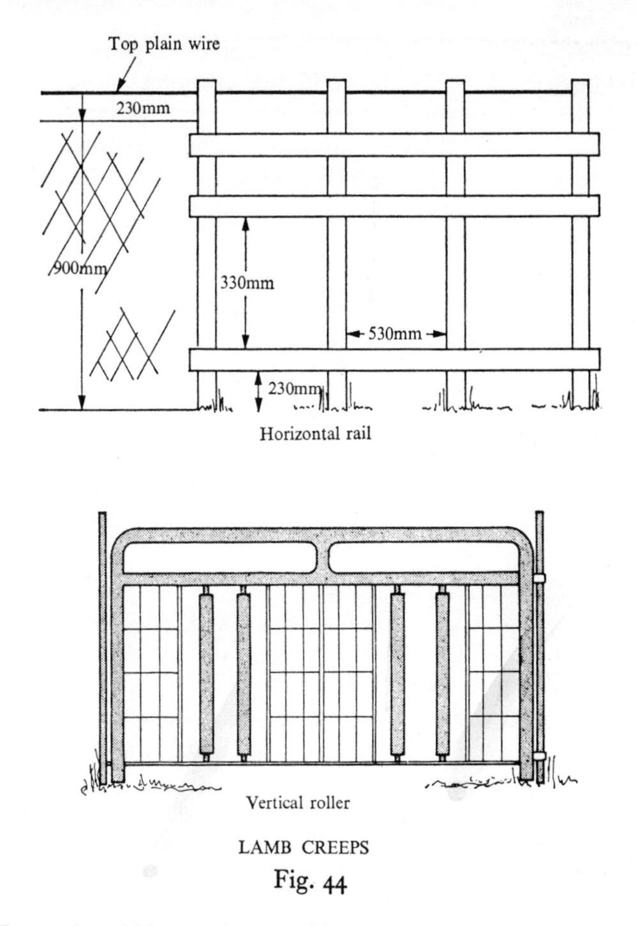

LAMB CREEPS

Fig. 44

Example: 120 couples at fifteen couples per hectare will require eight ha of pasture, which must be divided into six equal paddocks. Each paddock will be a little under 1·33 ha. This means that when the flock occupy a paddock there will be nearly ninety-two couples per ha.

The flock is moved every four to five days depending upon the amount of grass available. Should there be prolonged periods of rain it may be necessary to move the flock more frequently to avoid poaching the pasture, or to allow them an extra paddock, thus reducing the stocking density for a few days. Similarly in periods of drought when there is a shortage of grass it may be necessary to move the flock every one or two days.

One of the difficulties of the system is to get young lambs to leave their mothers and to enter the creep. Many ingenious devices have been tried, including placing logs, straw bales, milk crates, etc., in the creep area in order to arouse the lamb's curiosity. Occasionally a few ewes may be allowed in with the lambs to give the youngsters extra confidence.

Forward and sideways creep grazing offers the farmer a real opportunity to intensify sheep production; but, like most intensive farming systems, it demands an exceptionally high standard of husbandry if it is to be successful.

Dagging—crutching—burling—belting are all terms used for the removal of soiled wool from around the tail and anus. This task becomes necessary when sheep become laxative. It is particularly important in April and May since at this time of year the sheep carries a dense fleece, which increases the risk of blow-fly attack (see p. 139).

Weaning Lambs that are not sold as fat lambs must be weaned from the ewes during the late summer, and, traditionally, such lambs are at least four months old when weaned. The task is best done by taking the entire flock to a well-fenced field, with a good water supply. After two or three days the ewes are removed and taken to a bare pasture as far away as possible, preferably out of earshot.

The lambs will be restless and thirsty when weaned, but if the ewes had been with them for two or three days, they will have helped the lambs to settle and find the water tank. Never wean lambs abruptly by turning them into a strange field.

Work at the Grassland Research Institute has suggested that there is some advantage in weaning lambs at ten weeks old. This gives the lambs cleaner grazing which is claimed to offset the loss of ewe's milk. If the system is adopted, then the following points should be observed:

1 The lambs should be dosed against roundworms when weaned.

2 The lambs must be grazed on a rested, clean pasture, preferably a white clover-ryegrass mixture.

3 The ewes should be weaned on to the poorest pasture available to encourage 'drying off' the milk supply and to prevent their becoming fat before they are served in the autumn.

The advantage of ten-week weaning is that the lambs are free from competition with the ewes for grazing, and less likely to be infected with roundworms.

After weaning Inspect the flock regularly, keeping a watchful eye for overstocked udders, which should be milked out. Once the ewes have dried up, they may be changed to fresh pastures; but the aim should be to keep them down in condition until they are flushed in the autumn.

SUMMARY OF SHEPHERD'S CALENDAR

Making up the flock	Cull ewes with faulty udders and broken mouths, and check feet. Buy in sound ewes as replacements for culled ewes.
Purchase of breeding stock	Buy the best you can afford. Select sound, healthy ewes, good in foot, tooth and udder.
Flushing	Run flock over improved pasture to stimulate multiple ovulation.
Autumn dosing and vaccination	Dose flock for roundworms and liver fluke if necessary. Vaccinate yearling ewes with 5 cm³–2 cm³–2 cm³ system against clostridial disease.
Tupping	Allow 1 mature ram to 50–60 ewes. Use sire sine harness or raddle-change colour every 16–17 days. Start with yellow-red-blue. Remove ram after seven weeks.
Winter	Allow ewes to 'roam the farm', cleaning stubbles, leys or sugar-beet tops. Provide good quality, leafy, soft meadow hay from late November onwards.
Steaming up	Commence feeding 112 g concentrates per ewe per day, six weeks before lambing. Gradually increase concentrates up to 450 g per head just prior to lambing.
Lambing 'bag'	Make sure you have all the necessary supplies in hand well before the actual lambing time.

Vaccination	Inject all ewes with 2 cm^3 vaccine against clostridial disease two weeks before lambing, to give the lamb protection. The antibodies will pass to the lambs via the colostrum milk.
Lambing	Inspect flock every four hours. Ewes that lamb after 4 p.m. should be penned with their lambs for the night.
Castration and tailing	Elastrator and rubber rings should be used before lambs are forty-eight hours old. Burdizzo method should be carried out between 2–4 weeks.
Feeding	Allow up to 0·9 kg concentrates per head for ewes with single lambs. Ewes with twins should receive 1·3 kg concentrates per day.
Creep feeding	Lambs will start eating concentrates when two weeks old, and grass when three to four weeks. Fresh appetising concentrates should be offered *ad lib.* until lambs are eating 0·9 kg per day.
Routine worm drench	Dose all lambs when six weeks old, and monthly thereafter.
Grazing	Aim to change flock at least weekly on to new pastures. If possible, provide six paddocks, rotating flock every 4–5 days. Sheep prefer short dense swards with plenty of clovers.
Shearing	Remove fleece in one piece, avoiding second cuts. Roll tightly and pack in wool sack. Avoid contamination with string, oil, paint, and dirt. Store in a dry place.
Weaning	Place flock in well-fenced field with good water supply. Remove ewes from lambs after 2–3 days and put them on a bare pasture.
After weaning	Keep ewes in hard condition until the autumn.
Dipping	Immerse all sheep in an approved dip for at least one minute each or put them through a spray race.

Eight Sheep Handling

Types of sheep handling

Handling sheep is an important part of the shepherd's work. The best way to learn this job is to work with an experienced shepherd: watch how he moves slowly, yet deliberately, amongst his flock, and almost effortlessly will catch and turn up the individual he wishes to examine.

Catching

Pen your flock in a small enclosure, or, with the help of a sheep-dog, hold the sheep in the corner of a field. Move quietly into the flock, then quickly grasp the left hindleg firmly above the hock with your right hand. Slide your left hand along the shoulder under the neck and grasp the wool under the throat. Either halter the sheep or stand it against a fence, restraining it by placing your knee behind the shoulder.

If the sheep are penned tightly you will be able to catch them immediately by gripping the wool under the throat. Never hold the wool on any other part of the body or you will damage the fleece by breaking the wool fibres, and you may bruise the flesh, especially if handling young lambs.

A further way of catching sheep is to grip the loose skin found in the groin, and lift the sheep's hindleg off the ground.

Casting

There are innumerable ways of casting sheep. The aim should be surprise, firmness and gentleness, particularly if handling ewes that are in lamb. Ewes should only be turned up in an emergency during the latter stages of pregnancy.

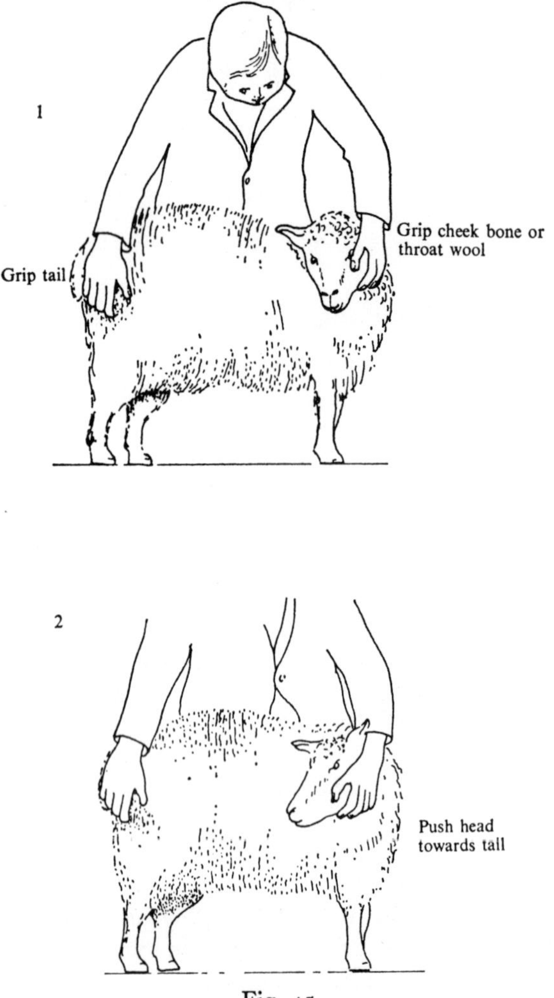

Fig. 45

1 Having caught your sheep, place your right hand under the tail whilst the left hand grips the throat wool; then gently push the head away from yourself and towards the tail. The sheep will roll over your knee and sit down comfortably. This is often rather difficult for the beginner to master, but certainly the quickest way for the experienced shepherd.

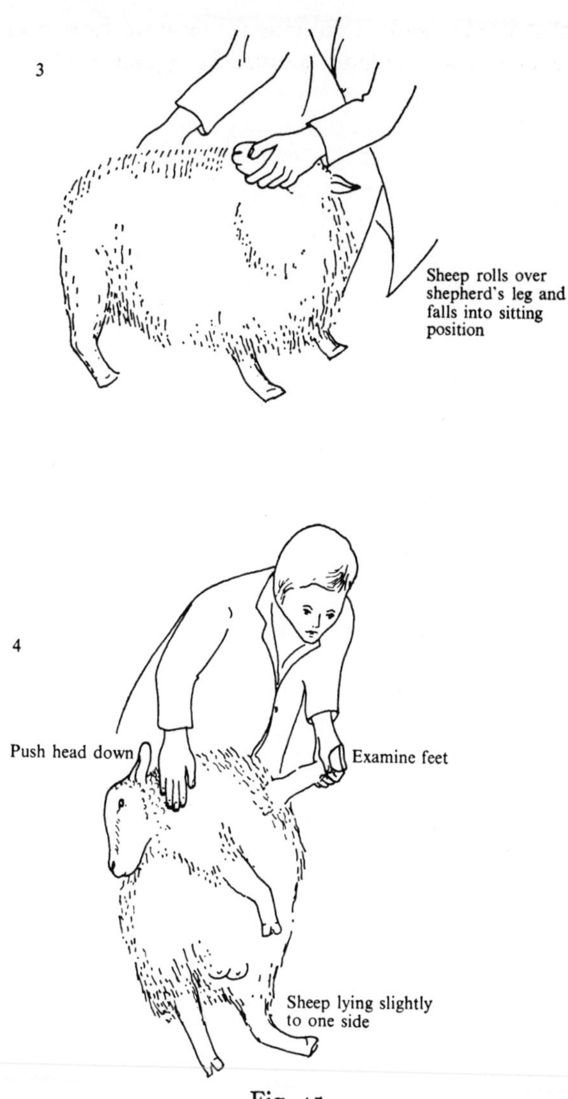

Fig. 45

2 Again hold the sheep by the throat wool with your left hand. Place your right arm under the belly and grip the offside hindleg above the hock. Pull the leg towards you, and the sheep will sit down. Quickly step round the sheep as it falls, then grip the sheep between your knees as it takes a sitting position.

Never allow a sheep to sit on its tail, but allow it to lie slightly to one side. This is an ideal method for heavy sheep and rams.

3 A similar method to the last is to place your right hand over the back and under the belly, grip the nearside hindleg above the hock, pull it away from you, and the sheep will fall into your lap. This method is more suited to small sheep.

4 Grip the throat wool with the left hand and place the right arm over the back and under the belly. Place your hand in a flat position across the belly, close to the nearside groin, then lift the sheep off the ground turning it away from yourself. Lower the sheep into a sitting position. This is the most common method of turning sheep, but is rather laborious.

The beginner should practise until he can safely and speedily turn up sheep, for it is an essential part of his work.

Dosing

Dosing sheep for protection against internal parasites is a routine practice on most farms. It requires careful handling, accurate preparation of dose and an exact method of administration. Either a drenching bottle or 'gun' may be used. The important points to remember are:

1 Always follow the maker's instructions when preparing the dose. Don't think that by exceeding the recommended dose that the drug will be more efficient—it will not, and in certain cases may prove fatal to the sheep.

2 If you are using an automatic dosing gun, check its accuracy before and during dosing by filling the gun and releasing contents into a graduated flask.

3 To dose the sheep, first restrain it in a drafting race, or hold it against a wall. Open the mouth by placing your left hand over the top jaw and thrusting your thumb into the side of the mouth. Then insert the dosing gun and work it towards the back of the throat, over the top of the tongue. When the sheep starts to 'chew' the gun gently force the contents down the oesophagus. With a dosing bottle you can rub the roof of the mouth, which will induce the sheep to salivate, and then you gently pour the dose down the throat. Stop immediately if the sheep chokes, or appears distressed.

Injections

There are three ways of injecting stock:

subcutaneous
intramuscular
intravenously

1 *Subcutaneous* is to inject a substance under the skin. The usual sites for this injection are over the chest wall, well behind the shoulder, or in the loose folds of skin found in the groin region. Behind the shoulder is most convenient, and a cleaner site with adult ewes, whilst with baby lambs the groin region is most suitable. The injection is done by lifting a fold of skin and thrusting the needle underneath, so that the needle point is in a pocket of skin. Gently release contents by forcing home the syringe plunger.

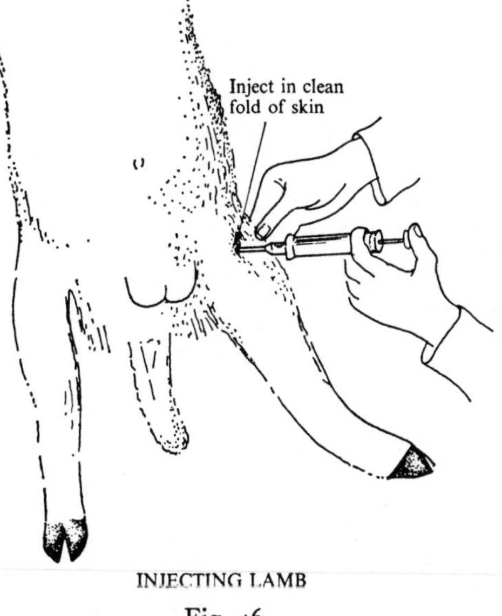

INJECTING LAMB

Fig. 46

2 *Intramuscular* is to inject directly into the muscle. The best site for this injection is in the fleshy part of the hindleg. Part the wool in a clean area, avoiding any soiled wool. Clean the skin with surgical spirit and plunge the needle straight into the muscle to make the injection.

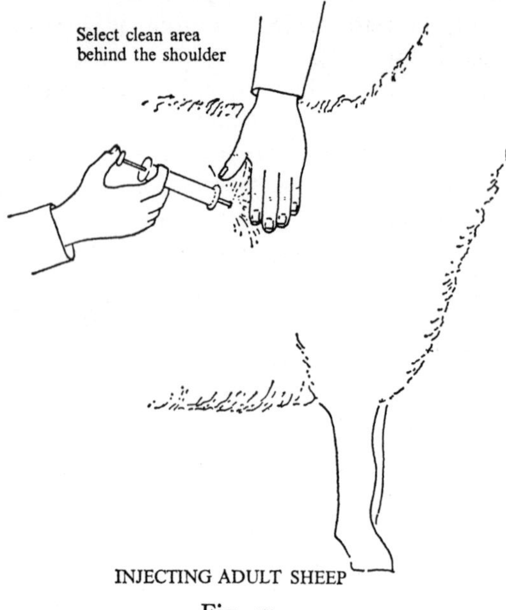

INJECTING ADULT SHEEP

Fig. 47

3 *Intravenous* injections should be carried out only by a veterinary surgeon.

Use of hypodermic syringe

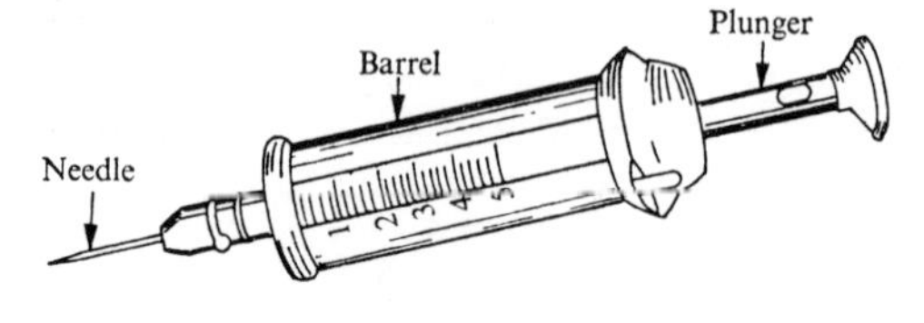

5cc HYPODERMIC SYRINGE

Fig. 48

Assemble the syringe and needle, shake the bottle and swab the cap with clean surgical spirit.

Draw into the syringe a volume of air slightly more than the volume of liquid to be withdrawn, and thrust the needle through the rubber cap of the bottle.

Turn the bottle upside down, push the plunger to inject the air in the syringe into the bottle. If you do not do this, you will have difficulty in withdrawing the dose, as a partial vacuum will form inside the bottle.

Make sure that the needle tip is well below the surface of the fluid.

Pull the plunger down, drawing slightly more liquid into the syringe than is required. Push the plunger slightly to expel any air bubbles, and adjust to the right dose.

Detach the syringe, leaving the needle in the cap for withdrawing subsequent doses.

Finally, attach a second needle to the syringe, expel any air from the needle, and make the injection.

Care of syringe and needles

Immediately after use the syringe should be dismantled, thoroughly cleansed, and then sterilised by boiling in clean water for twenty minutes. Needles should be changed between every 5–10 sheep, and then sterilised with the syringe. For sheep, a 12 mm or 15 mm 18 gauge is usually needed.

Note

1 Never vaccinate sheep in wet conditions. A wet fleece carries a much greater risk of infection.
2 Always discard part-used bottles at the end of the day.
3 Always check the dose on the bottle label, and make sure the correct amount will be delivered by the syringe.
4 Do not inject sheep within four weeks of slaughter.

Castration

The three methods of castrating lambs are
1 Elastrator and rubber ring
2 Burdizzo
3 Knife

1 *Elastrator and rubber rings* The rings must be applied before the lambs are two days old. The ring is placed over the scrotum and spermatic cord, immediately below the supernumerary teats. Care should be taken to see that both testicles are below the ring.

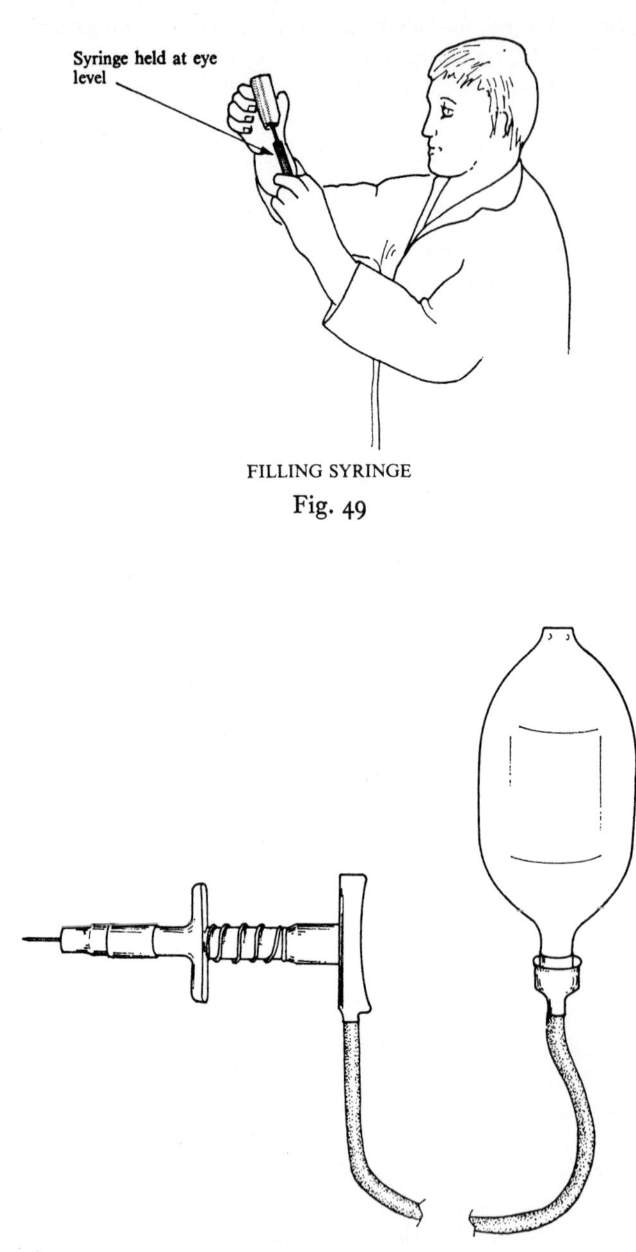

FILLING SYRINGE

Fig. 49

AUTOMATIC HYPODERMIC SYRINGE

Fig. 50

2 *Burdizzo or bloodless castrator* (medium or small size) Using a table, bench or bale of straw, have an assistant to hold the lamb as shown in Fig. 51.

Take the burdizzo in your right hand and carefully draw one testicle down with your left hand. Place the spermatic cord between the jaws, then apply pressure. Repeat with second testicle. Do not involve any more skin in the jaws than is necessary.

Lambs are best pinched before they are six weeks old.

Holding lamb in correct position for
(a) castration
(b) tailing
(c) ear-marking

Fig. 51

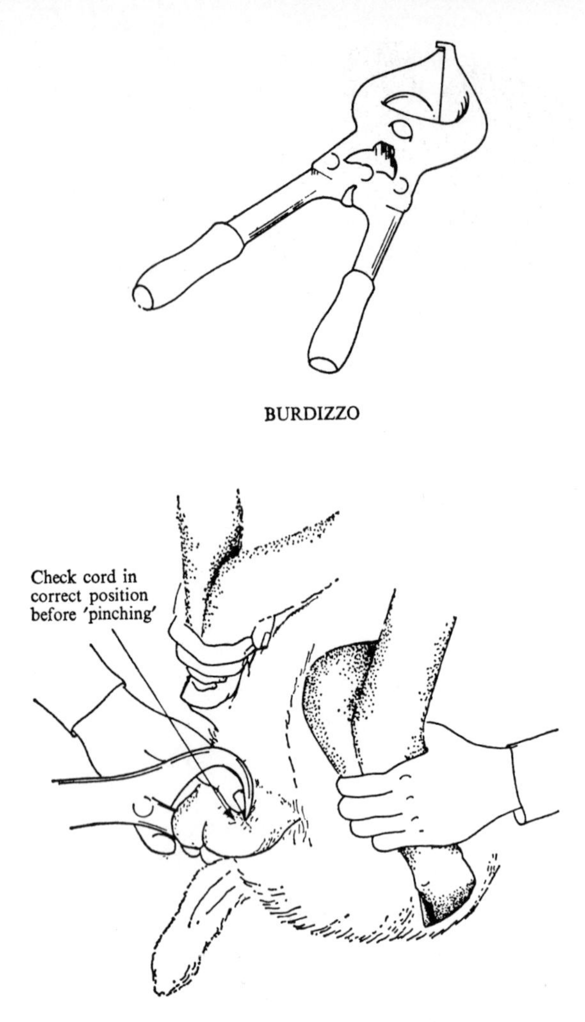

BURDIZZO

CASTRATION WITH A BURDIZZO

Fig. 52 (a) and (b)

3 *Knife or scalpel*

Equipment 1 bucket with clean warm water, soap, scalpel, 3 per cent Cetavolon

Note This is a surgical operation and should not be attempted unless accompanied by a veterinary surgeon or qualified teacher.

Wash hands thoroughly.

Wash the purse (scrotum) with warm water and Cetavolon or similar mild antiseptic.

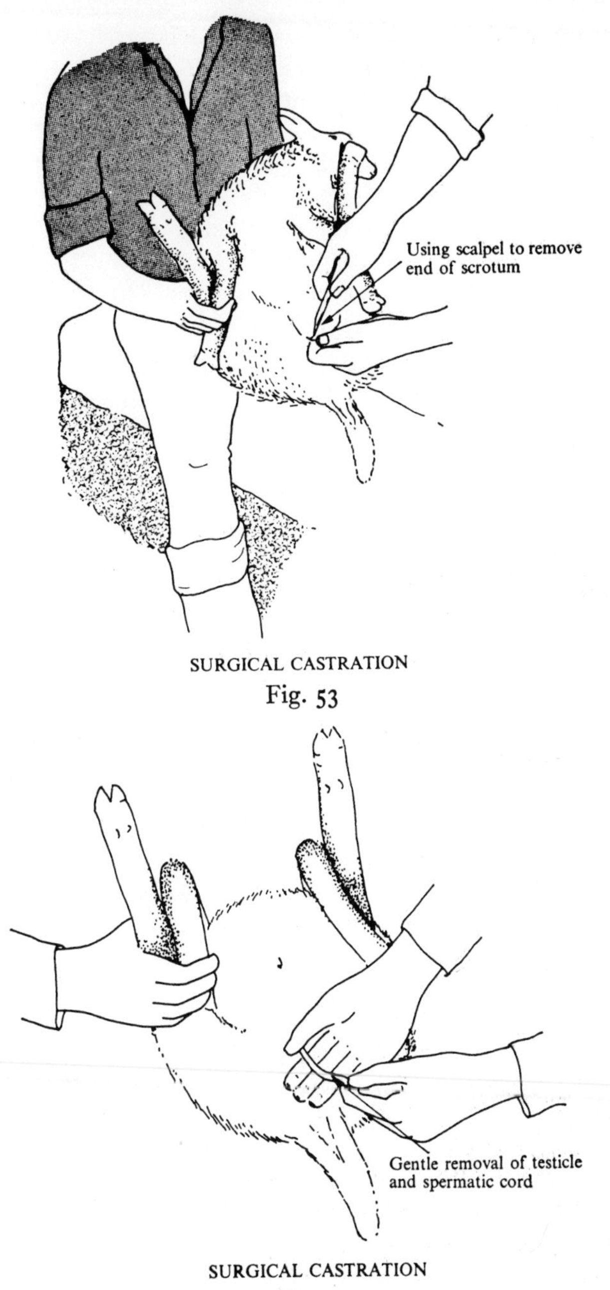

SURGICAL CASTRATION

Fig. 53

SURGICAL CASTRATION

Fig. 54

Grasp the end of the purse with the left hand and gently force testicles back with the right fingers.
Pick up sterile scalpel in right hand and cut off the end of the purse.
Push the skin back with the left fingers to expose the two testicles.
Grip a testicle between the thumb and first finger. Gently twist, and apply traction to remove the stone.
Repeat with second testicle.
Apply mild antiseptic to the wound.

Again it must stressed that these notes are given as a guide to students. You should always seek professional advice before attempting to castrate or dock an animal.

Tailing or docking

The four methods of tailing lambs are
1 Elastrator and rubber rings
2 Burdizzo and knife
3 Knife
4 Hot iron

1 *Elastrator and rubber rings* This is a fairly simple way of docking young lambs. The rubber ring is placed between the vertebrae joints of the tail, leaving about an inch of the tail, which is sufficient to cover the vulva in ewe lambs. In upland areas it is the practice to dock ewe lambs at hock height, so that the tail protects the udder. Rings should be applied before the lambs are forty-eight hours old. Normally the tail will drop off in 7–10 days.

2 *Burdizzo and knife* This is an effective way of tailing. The large-size burdizzo instrument is used to crush the tail between the vertebrae joints, then the tail is cut off with a knife. The crushing stops bleeding (see Fig. 56). Dust the wound with an antiseptic powder to prevent infection; this is most important when docking lambs in late May to avoid fly strike. Lambs are usually about two weeks old for this operation.

3 *Knife* Under favourable climatic conditions, lambs may be tailed with a sharp knife. The shepherd usually holds the lamb

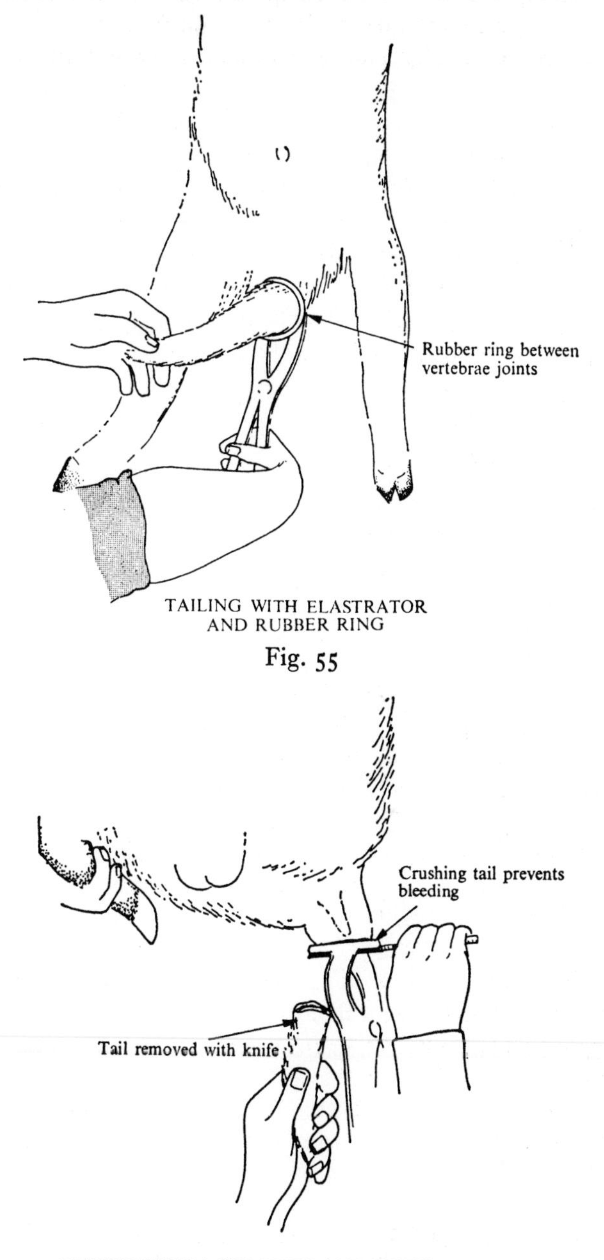

TAILING WITH ELASTRATOR
AND RUBBER RING

Fig. 55

TAILING WITH A BURDIZZO AND KNIFE

Fig. 56

between his knees, and then cuts off the tail between the vertebrae joints. There is some bleeding for a short time, but the lambs quickly recover. Lambs are best tailed when 7–10 days old by this method.

4 *Hot iron* This is less common today because of the problem of heating the iron. The shepherd makes a coal fire to heat the iron or uses a Primus stove. The lamb is held over a wooden table or the top rail of a gate. The hot iron is placed between the vertebrae joints to sever the tissue. The advantage of using a hot iron is that no bleeding occurs and there is little risk of infection.

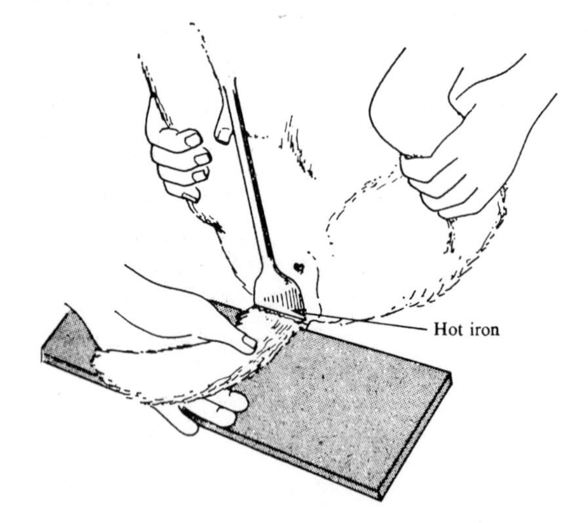

REMOVING TAIL WITH HOT IRON

Fig. 57

Nine Market Requirements and Sale of Fat Lamb and Mutton

There are two distinct types of fat lamb required by the meat trade. Firstly, the lightweight lamb that yields a compact carcass of 13 kg–16·6 kg, and, secondly, the larger carcass of 20·25 kg–22 kg which is suited to the needs of the catering trade. Unfortunately, no figures are available to show the exact requirements for each type of lamb, although one could generalise and say that there is less difficulty in selling small, lightweight lambs. Many farmers, however, argue that a heavy lamb will in fact make a bigger profit per head than a small lamb, although it makes less money per kilogram.

Butcher's lamb

The type of lamb most sought after by butchers today is the young (12–16 week) stocky, thick-set lamb with short, strong bone. These lambs yield a carcass with plenty of eye muscle and a light covering of white firm fat.

Suckling lambs sold straight from their mothers, for slaughter, are unlikely to be over-fat. Lambs that are weaned, allowed to go through a store period, and then fattened out as hoggets, require careful feeding otherwise they will produce carcasses with a small eye of muscle, and a heavy covering of fat. It is important, therefore, to feed the lambs on a high plane of nutrition—creep feeding, steaming up ewes to stimulate milk production, etc.—in order that they may grow as quickly as possible. Generally, we find that the best butchers' lambs are those sired by shortwool rams, e.g. Southdown, Ryeland, Suffolk, Dorset Down, etc., and are out of prolific, milky ewes such as Clun, Scots Half-bred or Masham.

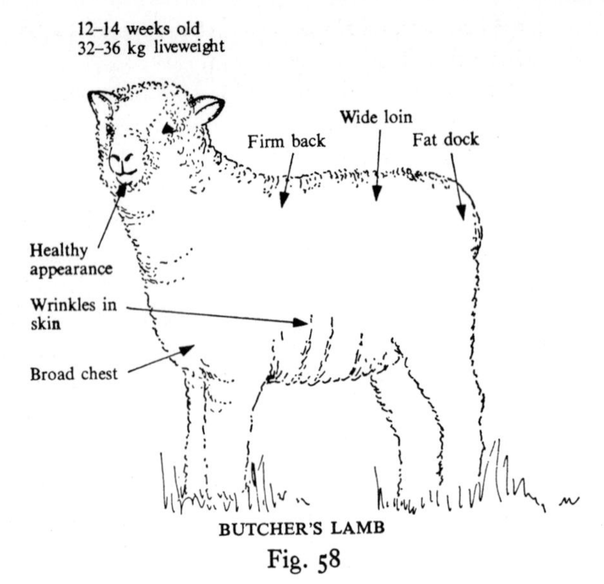

BUTCHER'S LAMB

Fig. 58

The carcass

The weight of a dressed carcass should be from 13 kg to 15·25 kg for the super-quality lamb; 15·25 kg to 19 kg for the good quality; and not exceeding 24·75 kg for mutton. The

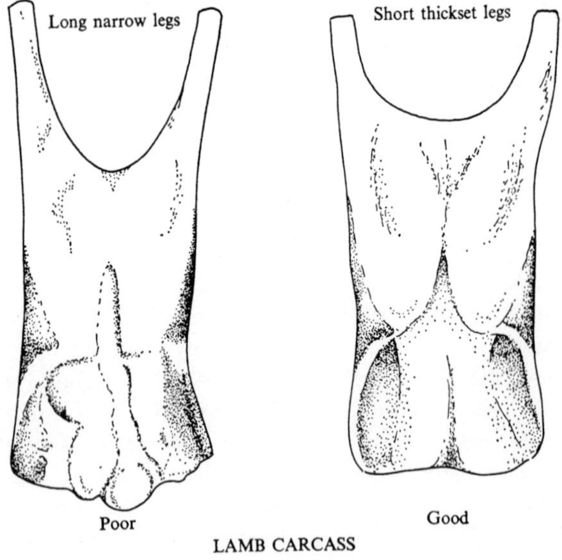

LAMB CARCASS

Fig. 59

carcass should be compact and relatively heavier in the hindquarters where the most expensive joints are found.

The leg must be well developed, the flesh being carried down to the hock. The loin must also be well developed and yield a deep eye of muscle when cut for chops, and the shoulder and neck should be light and free from excess fat.

Besides the carcass, the slaughtered lamb produces valuable edible offals namely, the liver, kidneys, tongue, heart, sweetbreads, head and caul, as well as inedible offals which the butcher sells to manufacturers—sheep skins are used for rugs or leather. The small intestines may make sausage skins, surgical stitches, or strings for musical instruments and tennis rackets.

Selection of lambs for marketing

Marketing lambs begin when the first of the crop are around 31 kg liveweight. It is possible, of course, to slaughter lambs at as light a weight as 14·8 kg liveweight, but this would be scarcely economical.

The lambs must be perfectly healthy and active, which will be apparent by looking closely at the eye which should be bright, and the fleece which should be tight. Next, one should handle the lambs to see if they have the right degree of finish. Place the palm of your hand over the loin, feel for muscle, and an adequate covering of firm flesh. Then handle along the back, and down to the tail or dock. The dock is a very good indication of finish in a lamb, and should give the impression of adequate muscle over the tail bones. Finally, one may feel the inside leg of a lamb, which, if it fills the hand and fleshes down to the hock joint, indicates a top-quality lamb.

Marketing

The lambs may be sold either:

1 On the hoof—in auction market
2 On the hook—on a deadweight basis

Either way the farmer will receive roughly the same price. However, if he wants to receive the government subsidy he must carefully obey the following rules:

Sale by auction—government guarantee

1 The guaranteed price for sheep applies to clean, fat lamb

and hoggets only. This means, of course, that male lambs must be properly castrated before they show ram-like tendencies, i.e. before six months old, and ewe lambs must be barren.

There is no subsidy for rams, ewes, pregnant ewe lambs, etc.

2 To qualify for payment, the lambs must have an estimated or actual dressed carcass weight of not less than 8·6 kg deadweight, and must be fairly well finished.

3 There is no maximum weight for sheep, but the guarantee payment is limited to the following actual, or estimated, dead carcass weight.

Lambs, 22·5 kg deadweight
Hoggets and other clean sheep, 27 kg

e.g. Fat lamb may be sold, having an estimated dead carcass weight of 28 kg, but any guaranteed payment that may be due will be calculated to 22·5 kg, the excess 5·5 kg receiving no government subsidy.

Arrangements for certification In order to obtain guarantee payment a producer must make application by filling in a prescribed postcard and sending it to the auctioneers seven days before he intends to market the animals. The auctioneer will then advertise the number of fat cattle, sheep, or pigs that will be available at the next fatstock auction in the local press.

Procedure at market On arrival at the auction market the sheep will be weighed, either singly or in level lots not exceeding twenty in number. Next, the sheep will be graded, and then offered by auction, according to their estimated dead carcass weight.

If the sheep are sold, before being moved from the market they will be permanently marked by a hole punched in the *left* ear.

Sale by deadweight basis

Lambs and hoggets may be sold to meat wholesalers and butchers, who will slaughter the stock and pay the producer an agreed price per kilogram deadweight. The carcass will be graded provided:

1 It is reasonably well fleshed throughout.

2 There is no sign of disease.

Note All meat is inspected by officers of local authorities before the carcass is sold for human consumption.

Payment Normally, the producer receives a cheque from the auctioneer, for the sale price, less the auctioneer's commission, within a week of the auction. The subsidy will be paid to the producer separately.

If the sheep are sold on a deadweight basis, the producer will receive an all-cash payment, which includes the subsidy payment.

Haulage and handling

It is most important that lambs are not handled roughly when they are being sorted for market or abattoir. Young lambs, especially, bruise easily and much damage can be done to the carcass if a lamb is roughly caught by its coat.

Always catch the lambs either by gripping the hind leg above the hock, and placing the free hand under the neck, or by gripping under the lamb's thigh (see Chapter Eight—sheep handling).

Carcass classification (see diagram opposite page 122)

The Meat and Livestock Commission operate a classification scheme, whereby carcasses are classified according to their weight, conformation, category and fat content.

Classification is run on a voluntary basis and helps both the meat producer wholesaler and retailers to evaluate the type of carcass required for home and European markets.

Codes

Carcasses are described by:

Weight as now defined for fatstock guarantee purposes. There is a detailed dressing specification which must be followed. The MLC check on weight and dressing specification ensures protection for the producer selling deadweight.

Category as defined for fatstock guarantee purposes: L–lamb, Hgt–hogget, Shp–sheep. Rams and ewes are not being classified at present.

Fatness determined by a visual appraisal of external fat development. There are five classes, ranging from 1 (very lean) to 5 (very fat).

In addition, the letter K is used to denote carcasses with excessive kidney knob and channel fat development.

The fatness of carcasses not only influences the cost of production, but has an impact on realisation values and consumer acceptance.

Conformation based on four classes: extra, average, poor and very poor. The conformation class is determined by a visual appraisal of shape, taking into account carcass thickness and blockiness and fullness of the legs. Fatness plays its part in influencing overall shape and no attempt is made to adjust for fatness.

Most carcasses fall in fat classes 2, 3 and 4. In these classes, carcasses of average conformation are identified by the fat class number only. Carcasses of very good conformation in these fat classes are identified by the letter E (Extra) following the fat class number. Any carcass of poor or very poor conformation, regardless of fat level, is classed as C or Z respectively and there is no subdivision between average and extra in fat classes 1 and 5.

Among carcasses of smiliar weight and fatness, good conformation carcasses tend to have greater lean meat thickness than those of poorer conformation. However, this does not generally provide a guide to the percentage of saleable meat in the side, nor of its distribution between high and lower valued cuts.

The sheep carcass classification scheme is, therefore, based on the ten class combinations shown in the following grid:

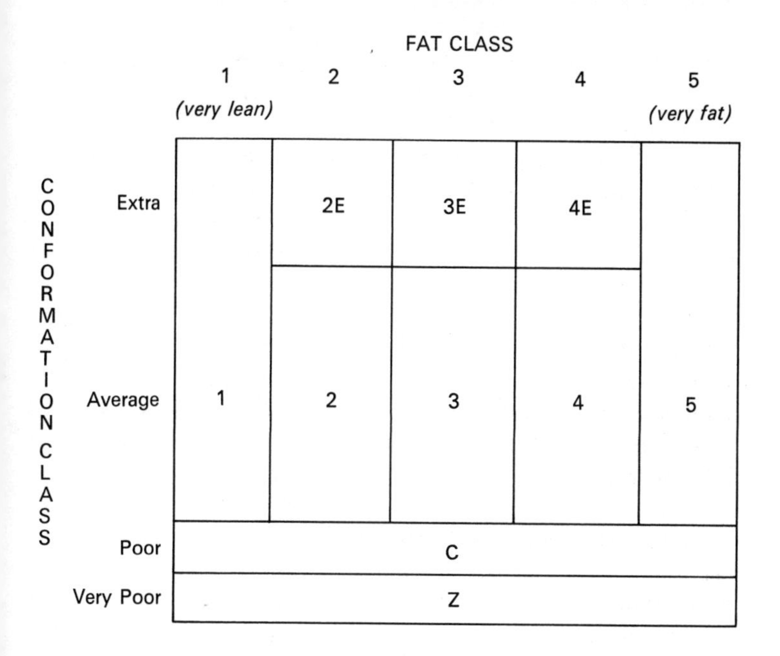

Ten Mutton and Lamb Production

The butcher's lamb is defined as a suckling lamb that is sold straight from its mother at between twelve and twenty weeks. Lambs that are weaned, allowed to pass through a store period, and then fattened are technically defined as mutton. However, mutton is an unfortunate word today, since many people associate the word with 'mutton fat' and heavy coarse carcasses. The butchery trade, almost universally, continues to call fat sheep 'lamb' until they are over one year old.

Store lamb

The autumn sheep sales or fairs play an important part in the shepherd's calendar. Thousands of lambs change hands at these auctions, and subsequent 'change of ground' has a beneficial effect upon the lambs.

The earliest sales are held in August and consist mainly of strong well-made Down cross lambs which will readily fatten in eight to ten weeks if grazed on good pastures. The later sales will be mainly hill lambs sold to the lowland farms for finishing.

The problem for the flock master is always to decide which lambs to buy, for, naturally, good lambs are expensive and poor quality lambs may be bought cheaply. Unfortunately, the best lambs, costing £20 or more, are not always the most profitable, particularly if they are marketed in late October–November. It may pay the farmer to buy second-quality lambs at around £15, keep them longer on the farm, and then market them after Christmas when prices are better.
are better.

Small but healthy hill sheep may be bought in late October–November and kept cheaply on the farm throughout the winter, then fattened the following spring on grass. These lambs are only given a little hay in severe winter, and roam the farm scavenging leys and stubbles in the winter months.

The lambs once sold to the fattener are now called hoggs or hoggets (see glossary) and they may be finished in one of several ways:

Root fed—arable farms
Grass fed—grass farms with supplementary concentrates
Barley lamb—an all-concentrate diet
Prime mutton—fattened at 12–18 months off grass or roots

Grass-fattening hoggets

Strong, healthy store lambs will readily fatten off good feeding pastures in September–October with little or no supplementary feed. Catch crops of rape and kale, broadcast after early potatoes have been harvested, may be also grazed as an extra feed.

Sheep are particularly fond of legumes, and do well on leys containing white clover or sainfoin pastures. Growth rate will naturally depend upon the quality of the herbage, but 0·9 kg–1·3 kg weekly may be expected under favourable conditions, the stronger lambs taking only eight to ten weeks to fatten. Slower-growing lambs will need some concentrates once the weather breaks. About 225 g of a mixture of equal parts rolled oats and barley will do as long as there is plenty of grass.

Any lambs that are not sold by the end of the grazing season should either be housed and fed as barley lamb or kept on a sheltered, dry field and fed good quality hay *ad lib.* and 0·9 kg of a suitable concentrate mix.

Either: 1 2 parts rolled oats
2 parts barley
2 parts flaked maize
1 part groundnut cake, plus vitamins and minerals

2 2 rolled oats or barley
2 flaked maize
1½ linseed cake, plus vitamins and minerals

3 3 rolled oats or barley
1 flaked maize
1 locust beans, plus vitamins and minerals

Root feeding

The majority of store lambs sold in September or October are fattened on root crops and root by-products, particularly sugar-beet tops. Other arable crops such as rape, kale and Brussels sprouts may also be fed. The lambs are folded over the crop to avoid wastage and to distribute the sheep's droppings evenly over the field.

With sugar beet tops, a rough guide is twenty-four lambs fattened per hectare, over a ten to fourteen-week period. The lambs are stocked at about ninety lambs per hectare, once the first part of the crop is lifted in October, and then folded over the fields until harvest is completed in December–January. The tops should be allowed to wilt for two weeks before feeding, because green tops contain oxalic acid. Ground chalk may be offered in self-feed hoppers as an additional prevention of 'stomach upsets'.

Good quality leafy hay must always be available for root-fed lambs, and a dry grass field or stubble should be accessible as a lie back should the weather become very wet.

Once frosty weather occurs, the tops will rapidly lose their feeding value, and some concentrates must be offered to keep the lambs growing. A suitable ration is:

2 parts rolled oats
2 parts rolled barley
1 part flaked maize
1 part groundnut cake
plus minerals

About 225 g per lamb of the above mixture will be adequate in the early stages of feeding, but as the weather and tops deteriorate you will have to increase the allowance to 450 g–675 g per lamb per day.

The best lambs should be ready for slaughter in December, although a few 'tail-enders' will need up to late January to finish. Some farmers grow a small hectarage of thousand headed and Hungry Gap kale for 'tail-end' lambs that have not reached market condition by the end of January. Other farmers will run the lambs on, and finish them in May off grass. These sheep will be twelve months old when slaughtered and are sold as prime mutton.

Prime mutton

A hundred years or so ago prime mutton was produced from three-year-old wether sheep. The carcasses were hung for about two weeks, and this helped to improve the tenderness. The meat was said to have excellent flavour! This would be scarcely economical today, and most of our mutton is produced as a secondary product from hoggets that, through reasons of ill health, poor management or extreme winter weather, were unable to be slaughtered at an earlier date.

The one advantage of yearling sheep is that they may be shorn in early May (about four weeks before shearing the lactating ewes). The fleeces are heavy and of high quality, and may be worth over £4 each, which the farmer receives in addition to the sale of the sheep.

Barley lambs

Since the advent of barley beef, many flock masters have considered the possibilities of barley lamb, or, in other words, intensive fattening of store lambs housed or kept in yards.

In the United States of America hundreds of thousands of lambs are finished each year on these lines, although maize and alfalfa hay is the standard feed rather than barley. (Maize is called corn in the U.S.A.) American experience of the intensive feeding of lambs suggests that lambs will make satisfactory liveweight gains if fed hay and corn at a ratio of 1:1, and that it is not advisable to exceed two parts corn to one of hay. Where lambs are fed larger amounts of corn their growth rate will be rapid (gains of up to 200 g per day have been recorded), but there is a considerable risk of lambs going 'off their feed' if forced too much. (Reference Kammalade and Kammalade.)

Experience in this country has shown that feeding rations containing up to 85 per cent barley with 15 per cent protein concentrate, plus vitamins and minerals, will give very good results. The lambs should be rationed with 225 g of good-quality hay and the concentrates offered *ad lib.* It must be stressed that to get good results you must feed strong, healthy, well-made lambs. The temptation to buy poor 'tail-end' lambs cheaply should be avoided. It is an advantage if the lambs are fed up to 450 g of concentrates per head per day before they are housed. This avoids the sudden change of diet from wet grass to dry meal. Before housing, pay particular attention to the feet, trim if necessary, and then run the lambs through a foot-rot bath containing 10 per cent copper sulphate.

It is advisable to vaccinate all the lambs with pulpy kidney vaccine and to dose them for worms. Wether lambs may be implanted with 30 mg hexoestrol, which may increase growth rate by 15–20 per cent and improve the food conversion.

Once housed, the lambs will quickly settle down to their main concentrate diet. Hay should be offered *ad lib.* for the first week to ten days, and then gradually reduced to 224 g–336 g per head per day. The concentrates are best offered in self-feed hoppers. Under good management the lambs will consume about 0·9 kg–1·3 kg concentrates daily, and a food conversion of 5:1 may be expected; if the ratio is markedly wider than this then a check on health must be made, and it may be necessary to dose the lambs again for roundworms.

If well-grown store lambs are bought weighing 26 kg–31 kg liveweight they will need about eight to ten weeks to fatten out at around 36 kg–40 kg.

Barley-fed lambs produce good carcasses with white firm fat and a useful proportion of eye muscle.

Seasonal supply of prime lamb and mutton

January February March	Canterbury lamb imported from New Zealand, a frozen carcass weighing 13 kg–16·6 kg—prime quality. English mutton produced from root-fed hoggs and 'barley lamb'.
April May	Easter lamb—prime English lamb produced from January-born lambs. Ewes fed mainly on concentrates and hay. Some supplies of mutton from wether sheep fattened on the spring flush of grass.

June July August	Prime English lamb produced from March-born lambs. Ewes fed grass only.
September October November	Hoggets fattened on grass with some supplementary concentrates and hay.
Late November December	Root-fattened hoggets and barley lambs.

Fig. 60 **Forage crops available for summer feeding of lambs**

Type	*Season of use*	*Yield fresh weight per hectare* (tonnes)	*Dry matter yield per hectare* (tonnes)	*Days feeding for lambs*
Continental/ stubble turnips	July–September	86 (70–99)*	7·7	99 lambs/ha for 6–8 weeks (4900 lamb feeding days/ha)
Fodder radish	August–October	37 (25–44)	5·2	50–60 lambs/ha for 6 weeks (2350 lamb feeding days/ha)
Kale	August–October	44 (29–50)	6·2	60–70 lambs/ha for 6 weeks (2790 lamb feeding days/ha)
Forage rape	July–October	32 (17–40)	4·5	60 lambs/ha for 6–8 weeks (2700 lamb feeding days/ha)

* Typical range

Source: MLC

Fig. 61 **Forage crops available for autumn and winter feeding of lambs**

Type	*Season of use*	*Yield fresh weight per hectare* (tonnes)	*Dry matter yield per hectare* (tonnes)	*Days feeding for lambs*
Continental/ stubble turnips	September–December	37 (30–60)*	3·4	60 lambs/ha for 6 weeks (2500 lamb feeding days/ha)
Forage rape	September–November	28 (16–37)	3·9	50–60 lambs/ha for 6 weeks (2300 lamb feeding days/ha)
Sugar beet tops	October–November	17 (9–24)	2·7	35–40 lambs/ha for 6–8 weeks (1800 lamb feeding days/ha)
White fleshed turnips	October–November	66 (50–85)	5·9	50–60 lambs/ha for 6–8 weeks (2700 lamb feeding days/ha)
Kale	September–November	35 (25–40)	4·9	50–60 lambs/ha for 6 weeks (2300 lamb feeding days/ha)
Yellow fleshed turnips	November–January	60 (44–74)	6·0	50–60 lambs/ha for 6–8 weeks (2700 lamb feeding days/ha)
Winter cabbage	November–January	85 (61–98)	12·7	70–80 lambs/ha for 8 weeks (4200 lamb feeding days/ha)
Swedes	November–March	60 (44–74)	7·2	75–95 lambs/ha for 12–14 weeks (7700 lamb feeding days/ha)

* Typical range

Source: MLC

Eleven Selection, Trimming and Showing Sheep

The agricultural show is an important 'shop window' for the farming industry; it is a meeting point where breeders from home and overseas can get together to discuss and inspect the very best British livestock. Although it may be argued that certain classes at shows are a trifle outdated, the local and national shows still command considerable interest and respect from both the farming community and the general public.

To be able to exhibit stock successfully a great deal of knowledge, skill, patience, and hard work are needed. The job involves first selecting a number of suitable animals, and then carefully feeding and managing them in such a way that they reach their peak condition for the show date. Of course, the final results, if successful, are extremely rewarding, but remember there is also disappointment, for in every show ring someone has to lose.

Selection of stock (see also p. 63)

Only the very best animals should be chosen for preparing for a show. Exhibiting second rate stock is not only detrimental to your reputation, but will detract from the image of the breed you represent in the minds of fellow breeders and prospective breeders. Generally speaking, we find that the successful animals are those that grow and improve steadily, without any check, from the time they are born until the show day.

In selecting sheep for show purposes, look first for the important characteristics, such as growth rate, conformation, wool quality and breed character. In general we look for the butcher's characteristics of the low-set, thick, compact, deeply muscled sheep with good spring of ribs, and plenty of heart room. The outline must have a certain smoothness, as well as balance and style. Breed character is important to a certain extent, but should never be placed over the basic commercial qualities.

Trimming equipment

The art of trimming the fleece for the show ring cannot be mastered from studying books alone. It calls for persistent practice, and where possible should be aided by the guidance of an experienced shepherd. However, if you have no one to help you, practise with some fattening tegs just before you send them to slaughter.

You will need to buy certain equipment, and, with trimming shears especially, it is well worth while buying the very best obtainable. Besides the shears, you will need a coarse and fine wool card, a flatboard, dandy brush, several halters, an oil stone, and an old bent eating fork for raking wool out of the card.

Trimming

Tie the sheep up with a halter, or fasten it in a Y stand with the head secured in the normal position.

Start by brushing the fleece with a dandy brush to remove any bits of loose straw, soil or other foreign matter. Damp the wool lightly with a mild solution of sheep dip (about half a cup of dip in nine litres of water) by dipping the brush into the liquid and then shaking the brush over the fleece.

Next take the coarse wool card and place firmly in the fleece, break the wool out by levering the card out of the wool in a backwards, revolving motion. Be careful not to dig the card in too deeply to begin with, for this will cause discomfort to the sheep. Gently, but firmly, card out all the wool, commencing at the base of the neck, then proceeding along the back, and finally under the neck and along each side. You will find a bent fork useful to clean out the wool in the card from time to time.

Using the shears

Before starting to trim, stand well back from the sheep and observe its outline. Decide in which way you are going to trim, and form a picture in your mind's eye of the end product you wish to achieve. Most shepherds begin by trimming the back, and finish up around the breech area. There are two methods of using the shear: you can either close the blades against each other as in closing scissors, or, by holding the one blade still, draw the other blade over the first shear. The latter method is perhaps the more difficult to acquire, but once mastered it will prove to be less tiring on the hands, and the resulting work will be much smoother.

As trimming progresses, the bottom blade moves along the fleece, away from the operator, with the top blade cutting the tips of the wool. The bottom blade should be pressed slightly into the fleece to give an even, level cut.

Make sure that your sheep stands firmly on all four legs throughout the trimming, or you will finish up with uneven patches of wool. Once you have 'backed down' your sheep sufficiently, you will have formed the outer contour of the sides. Hard backing, that is trimming the wool down to about 6 mm over the spine, will give the sheep a wide level back which will exaggerate the bulkiness of your sheep.

The sides are comparatively easy to trim. All that is needed is to follow the body contour, after first carding the wool out fully. A board is useful at this stage to pat the ends of the wool, and give fulness and firmness to the fleece. Trim the neck and breast to give boldness in rams and femininity in ewes. This is achieved by heavy carding with rams to give extra fullness after a light trimming. Ewes are not carded so much and perhaps trimmed a little harder to bring out the feminine characteristics. The rump, thighs and dock need extra care if the desired results are to be achieved. The dock may be trimmed level or a slight curve given. The thighs must be carded out to give extra fullness.

Finally, you may either trim the head or leave it until nearer the show. In fact, many shepherds trim the head on the show-ground, in order that the wool does not inadvertently become soiled or damaged before the show date. Skilful trimming of the head will emphasise the breed character, length or shortness of ears, wool around the eyes, and clearness of face. Blackface sheep, such as the Suffolk, may have their faces and ears and legs lightly dressed with a vegetable oil to give an improved 'bloom'.

Dressing

Occasionally you may find a sheep with extremely dry wool (little yolk). This can be overcome by lightly oiling the fleece with lanolin (wool fat) or a suitable mineral or vegetable oil, such as liquid paraffin or olive oil. The best way to apply the oil is to dip your fingers in it and then pat the wool with your hands. It is sometimes necessary to 'oil' fleeces several times before a show, at weekly intervals.

Blanketing

After trimming, blanketing is recommended, for not only does it keep the dirt out but it also acts as a conditioner. The wool improves in 'feel' and density. Blankets can be made quite easily from cotton 'calf milk' sacks, or hessian bags, or canvas. The important thing is that the blanket fits snugly and that no straps pinch or chafe the sheep.

Show day

The sheep should be lightly carded and trimmed again either the night before or on the morning of the show. The final appearance will be improved if the entire fleece is lightly patted with the carder and then the board. This will also remove any marks made by the blanket.

Showing sheep

Sheep must be trained to lead on a halter well in advance of the show season. Young animals will quickly respond to training, especially if an older sheep is walked in front of them. He must become accustomed to handling and not jump away when the judge feels his back or dock. He should also be trained to stand well, with feet wide apart, and set squarely under each corner.

Make every effort to get your stock to the showground in good time, in order that they may relax before being shown. A nervous or tense animal will be 'pinched up', and not show to full advantage.

Lead your stock slowly and deliberately around the ring. Remember that it is as important for the sheep to see the judge as for the judge to look at the sheep. In other words, keep out of the way yourself, and allow the judge and sheep to enjoy each other's company.

When you are invited to bring your exhibit in for handling, line your sheep up squarely with the others, and squat down by the sheep's side. You may then press the sheep back into a good standing position by pressing the sheep's breast with your knee. If your sheep is sagging in the middle, gently tap him under the belly, so that he straightens his back.

If the judge asks you to move your sheep up or down the line, do so quietly, always moving them into the new position from behind the line. Remember to show your animal from the moment you enter the ring until you leave, and try to smile just as much if you lose as if you win.

Twelve Sheep Shearing

Sheep shearing is a most difficult skill to acquire, and considerable study and practise are necessary in order to become proficient in the art. Shearing is very hard work, often performed in extremely hot and unpleasant conditions. In order to succeed, it is most important that the learner possesses the ability to work hard and the determination to become a good shearer.

The beginner must learn how to catch, turn up, and control the sheep whilst it is shorn. He must study shearing techniques, and fully master the correct use of the shearing handpiece. It takes at least two English seasons and the shearing of several hundred sheep for the novice to become thoroughly competent in the basic principles of shearing. During this time he should concentrate on the proper control of his sheep, and on shearing cleanly.

On no account should he try to become a fast or 'gun' shearer too early in his career, for the temptation to shear quickly leads to rough careless work.

To shear well means to remove the entire fleece in one piece, with no 'second cuts' in the wool, as quickly as possible with the minimum of discomfort to the sheep. 'Second cuts' means the action of shearing the same area twice. If the handpiece is allowed to rise off the sheep's skin it will cut through the staple of wool making it necessary to shear the half-cut area again, thus calling for a second cut. These reduce the wool to little more than chaff, and so reduce its value considerably.

Speed

Beginners should avoid the temptation to try to shear fast at the expense of good workmanship. If the learner can shear a ewe in 7–10 minutes in his first season this is satisfactory. Fast shearing comes with experience in handling the sheep, avoiding second cuts, making quick returns at the end of each blow, and keeping the gear in tip-top order.

Professional sheep shearers

In this country, contract shearers aim at clipping 150–200 sheep per day, and they are paid on a headage basis of around 30p per sheep. The contractor provides his own equipment; the farmer employs a man to catch the sheep, and wind the wool.

In New Zealand and Australia 'gun shearers' clip upwards of three hundred sheep, working a ten-hour day. Mr. Godfrey Bowen, the world-famous New Zealand shearer, clipped 456 sheep in one day in 1953. The total weight of wool shorn was approximately 1800 kg. The record was later broken by Mr. Brian Davies of Sennybridge, Powys who in 1976 clipped 609 sheep in nine hours. Two weeks later a fellow Powys farmer, Mr. W. Phillips, clipped 690 sheep—the present world record.

Time of year

Sheep shear best when the weather is warm and sunny. The wool is said to 'rise' when the sebaceous sweat glands produce a yellow-coloured fluid, known amongst shepherds as the 'yolk'. In Southern England the weather is often favourable for shearing in late May, though the majority of sheep are shorn in June. In upland areas and the North of Scotland shearing may be delayed until late July.

Weather hazards

On no account should sheep be shorn under wet conditions, for this may cause severe losses. Lactating ewes are particularly susceptible to chills and mastitis in the udder if shorn in wet or cold weather. In extreme cases ewes have died, owing to the sudden stress of losing their protective coat under bad weather conditions. It is, of course, possible for the sheep owner to be over-cautious and leave shearing until too late in the season.

Once favourable weather conditions prevail, shearing should proceed without delay. This will aiso avoid damage from the blow-fly.

Blow-fly attack

In warm, humid conditions blow-flies lay their eggs in sheep's wool, usually around the tail and over the back. Dirty and urine-stained wool is particularly attractive to the flies. Within 24–48 hours the eggs change into larvae, which burrow into the skin, causing extreme pain and discomfort to the sheep.

Preparation for shearing

Dress

The shearer should wear comfortable old clothes for shearing. A woollen singlet or an old sleeveless pullover will allow freedom of movement, and, at the same time, absorb perspiration. Woollen trousers will absorb much of the grease out of the fleece and prevent it seeping on to the shearer's legs. The wool grease is slightly acid and may cause a skin irritation or grease boils if the shearer has a sensitive skin.

Footwear

Light footwear such as old canvas shoes, casuals, or moccasins are ideal for shearing. Gumboots and heavy, hob-nailed boots should not be worn, for the shearer's feet play an important part in controlling the sheep and a thin sole enables him to anticipate the struggles of a restless sheep.

Shearing equipment

Fig. 63 shows a complete shearing unit. The 0·2 kw electric motor is suspended over the floor at such a height as to allow the flexible drive, which is called 'the dropper', just to clear or 'kiss the floor'. Thus, when the shearer takes the handpiece back at the end of each blow, a pendulum action carries the hand-

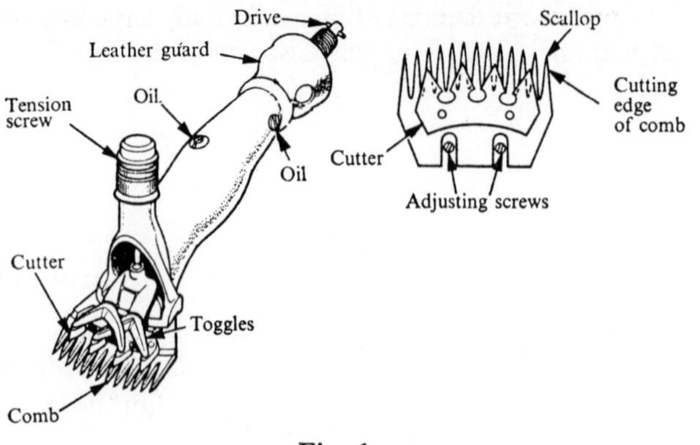

Fig. 62

piece forward. If the motor is set too low there will be too much play on the dropper, which will allow this part to get tangled up behind the operator.

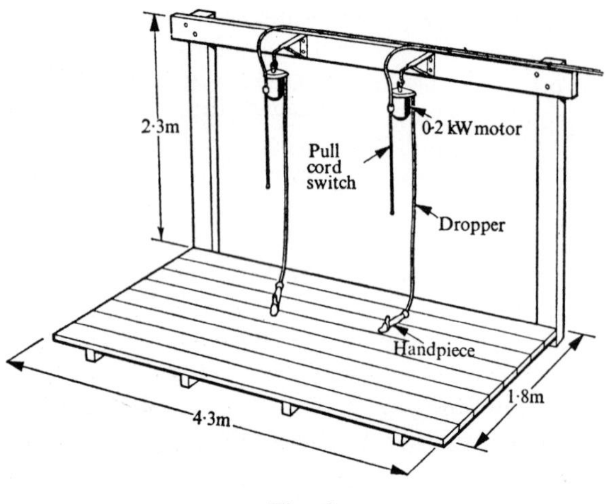

Fig. 63

The electric motor must be correctly wired and connected to a reliable electric system, with, of course, an *earth connection. The handpiece* is a precision-built machine, and will give a lifetime of service if properly maintained.

In setting up the handpiece for use, four important points must be kept in mind:

1 Lead
2 Tension
3 Throw
4 Lubrication

Lead This is the distance between the points of the cutter and the scallop on the comb. For shearing early in the season, or for shearing dirty sheep or lambs, no lead is used, the cutter tip being set up to the edge of the scallop on the comb.

As shearing conditions improve, and the grease rises in the wool, lead is introduced. The comb is pushed forward on the adjustment screws to give a 5 mm distance between the tip of the cutter and the edge of the scallop. The increased lead will avoid the risk of cutting the sheep's skin because the increased distance between the tip of the comb and the tip of the cutter allows the comb to ride over wrinkles in the sheep's skin. Wrinkles are more pronounced under good shearing conditions when the weather is hot.

Tension on the cutter and comb There are two mistakes that the beginner is likely to make in setting the tension on the cutter. The first is to give too little tension, which allows dirt to get between the cutter and comb, thus quickly blunting the gear. The second is to give too much tension, which causes overheating of the metal.

Correct tension is found by turning the tension screw, one notch at a time, until it is just possible to reciprocate the cutter by locking one's thumb in the rear cog of the handpiece (disconnected from the dropper) and rotating the drive. This is a guide only, and the tension will depend upon the strength of the shearer's thumb. However, if a ragged cut is being made as one starts to shear, the tension can be increased a notch or two.

However, if a perfectly clean cut is effected, then the tension should be slackened a notch and the cutter will probably clip four or five extra sheep.

Throw is the distance that the cutter travels from side to side of the comb. Ideally the cutter should meet each side of the comb evenly.

If a new cutter is used with an old comb it may be found that the cutter travels over the side of the comb and tears the uncut wool.

By moving the comb at the adjusting screws the throw can be transferred to the side of the comb that is farthest away from the fleece. In this position the cutter will overthrow on the off side and not damage the wool.

Care of combs Properly cared-for combs will last for several years. It could be argued that a comb improves with grinding; the professional shearer will use his old combs early in the year, because from continued sharpening they are thinner and therefore penetrate the wool easier.

After use the comb should be washed in water, and then dipped in oil to protect the metal against rust.

New combs The points of new combs should be carefully examined before use. If they are sharp and pointed it is worth while rubbing their tips with emery paper, and finally polishing them by drawing the comb through a piece of soft wood.

This polishing effect will prevent the new comb 'dragging' in the wool when shearing.

Lubrication The barrel of the handpiece should be filled twice daily with 30-grade oil. The drive cogs, tension screw and the cutter and comb should receive a light coating of oil at least every half-hour, and under difficult conditions between each sheep shorn.

If the handpiece becomes hot, examine the tension on the comb and cutter, and then look at the rear-drive cogs. It may be that wool has penetrated the leather guard and wound itself on to the gears, or the friction may be due to a lack of oil.

A pictorial guide to high-speed sheep shearing

There is more to shearing than the sheep and the shears. As a would-be expert you will need practice, patience and perseverance in full measure. Study these illustrations carefully (the numbers in the drawings show the sequence of blows in each position) and attend what demonstrations you can.

Catch your sheep by standing behind it with your feet either side of the sheep's hindfeet. Hook an arm under its throat.

With a smart, upward lift of the arm the sheep's fore-end is raised off the floor. Walk back quickly and the sheep will move its hind legs and will walk of its own accord.

By this method, the sheep arrives on the board in the right position for shearing*

1. Sheep slumped so brisket is 150 mm lower than if sheep sat bolt upright. Shearer's left arm brushes brisket wool up, left hand clasped near front leg. Stretch sheep up and slightly backward to remove wrinkle when shearing paunch.

*From the *Farmer and Stockbreeder*, 9 April 1968 (Drawings by Ted Fellows)

2. During this sequence tuck near front leg behind knee. Left hand presses sheep's offside foreleg to brisket and fingers stretch belly skin for shearing.

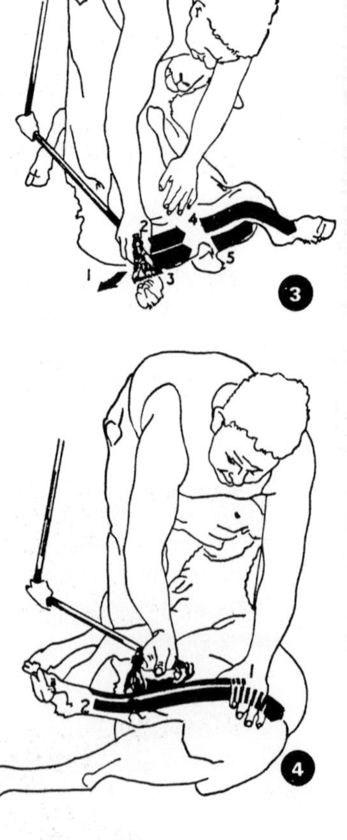

3. From finish of belly, run out near hind-leg, then two blows near-side crutch. Fingers of left hand guard teats. Fourth blow run out inside off leg. Fifth blow out on off leg to complete crutch.

4. Left hand in flank to straighten leg. First blow out the top of leg to part the wool for next blow which turns and comes back into the flank.

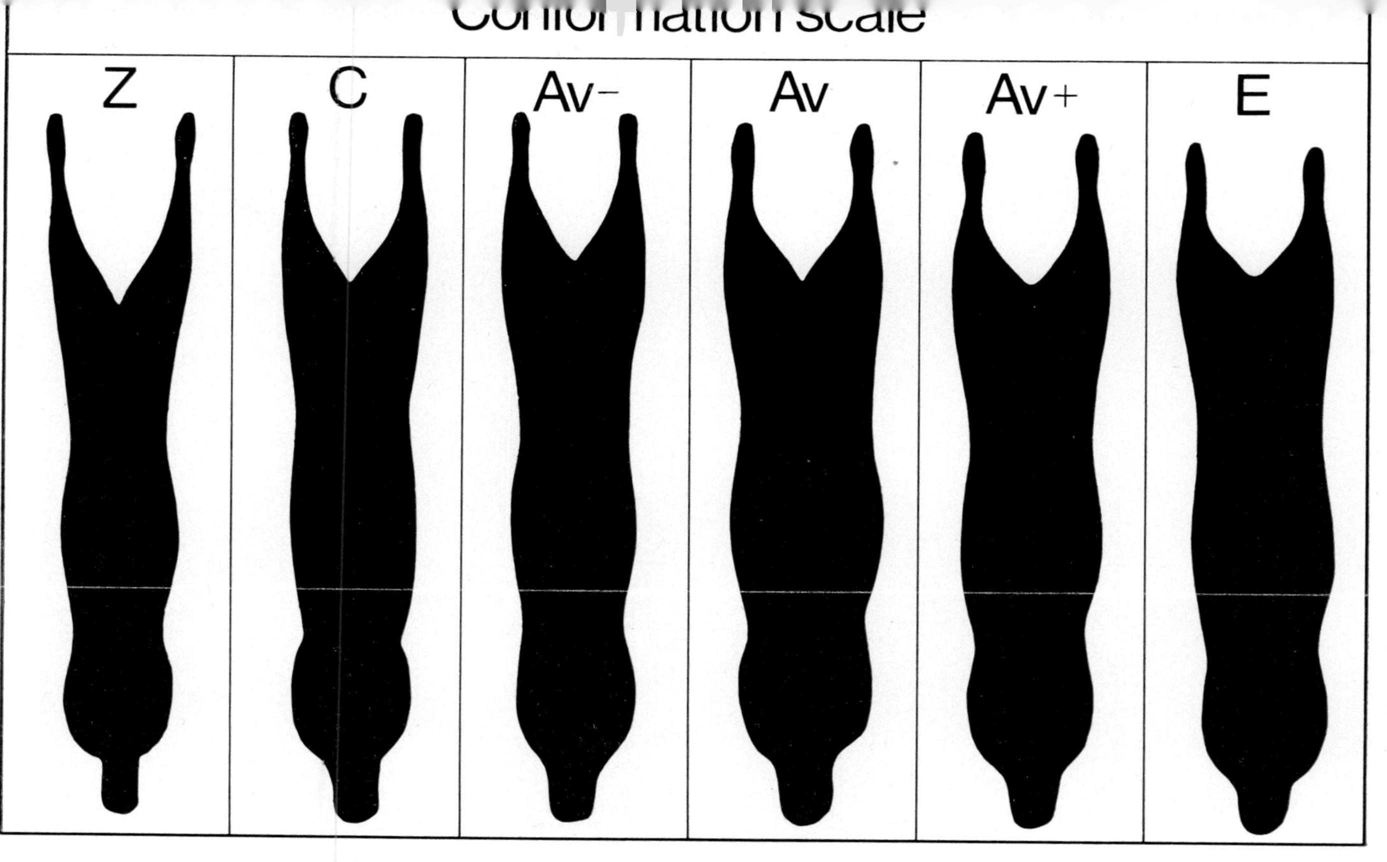

MLC Sheep carcass classification scheme

Ashford Rival—Champion Ryeland Ram—shown here by his breeder Mr. Robert Webb, illustrates all the care and skill

Trimming equipment, l–r: dandy brush—halter—shears—wool card—old bent fork—flat board—coarse comb

Halter lamb and tie up to heavy trough—observe the general outline and contour of sheep

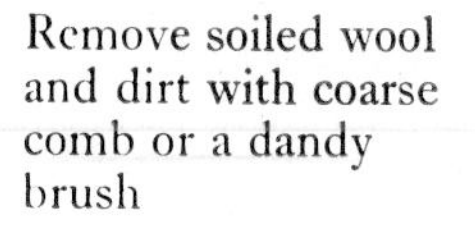

Remove soiled wool and dirt with coarse comb or a dandy brush

Using the wool carder, break out the wool gently

Dampen the fleece with a mild solution of sheep dip

Trim the hind leg

Trim the back to form a flat table-like appearance

Complete the tail area and hind leg

Trim the breast to give fullness

Lightly trim the topknot and face

After dampening the coat firm the wool by tapping gently with a flat board

The result after the first trim —at least two more trimmings will be required before the sheep is shown

Feeding time inside the timber shed where telegraph poles and straw bales provide a cheap overwintering building (Photo © Keith Huggett; reproduced by permission of *Farmer's Weekly*)

Sheep severely infested with the scab parasite (Photo: 'Farming Outlook', Tyne Tees Television)

5. First blow from hock straight down diagonally almost to backbone. Comb kept on firm part of leg to paunch. Leg shorn to natural pattern. At end of blows, right hand and wrist tip out to keep comb on sheep.

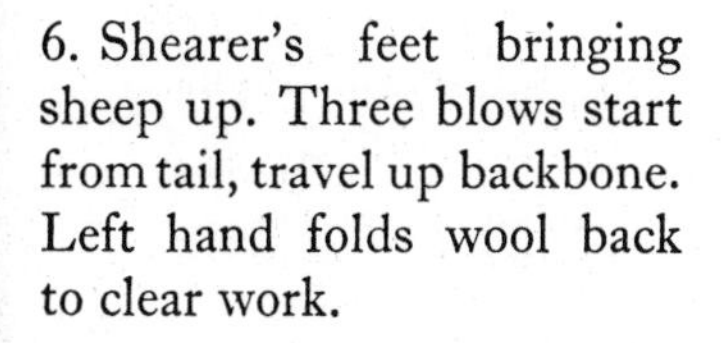

6. Shearer's feet bringing sheep up. Three blows start from tail, travel up backbone. Left hand folds wool back to clear work.

7. From over the tail the sheep moves up by shift of shearer's legs, top-knot removed as sheep moves.

8. Start neck on brisket. Left hand stretches wool back. Shearer's left leg stepped well through backbone, sheep leaning out round left thigh and knee. Shearer's right leg between sheep's legs.

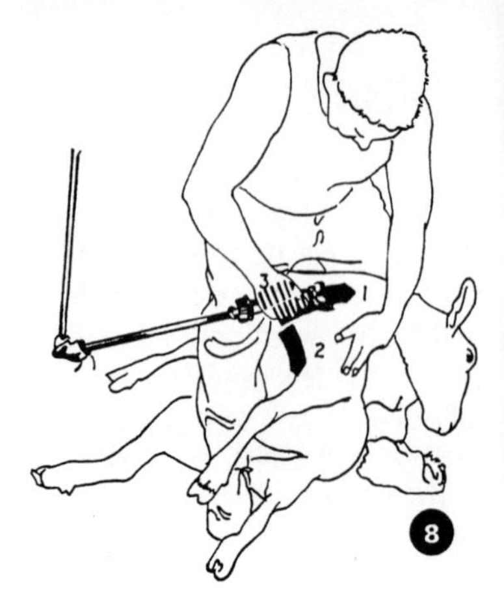

9. Straight up the neck under the throat. Wool broken out with outward turn of handpiece.

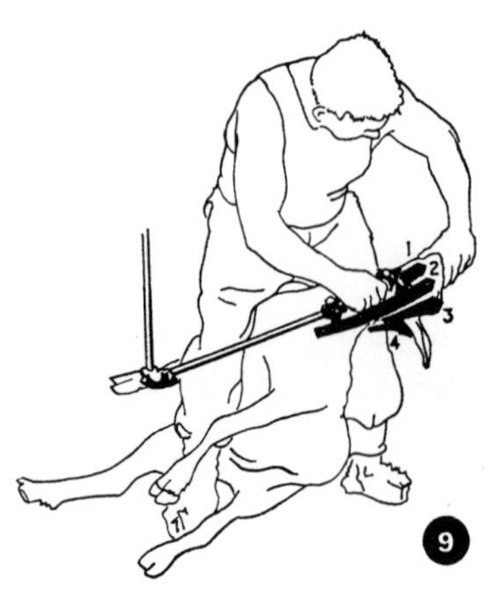

10. Take ear in left hand which rolls head on left knee. One or two blows across poll to other ear.

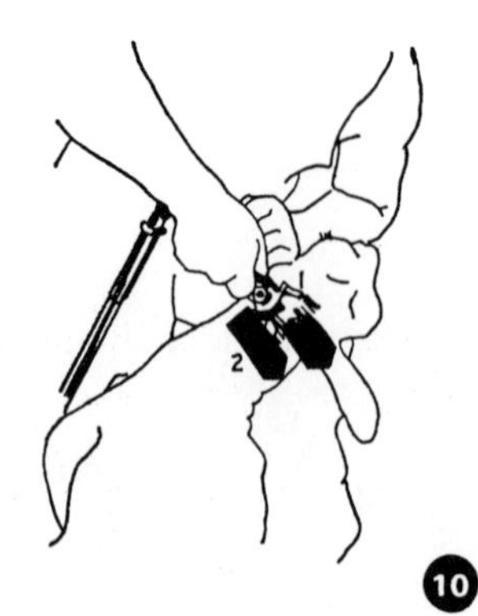

11. Following blows come in from the off front leg and clear the point of shoulder. Elbow of left arm holds sheep's head against the knee. Shearer's feet moving few inches at time brings sheep around to the long blow position.

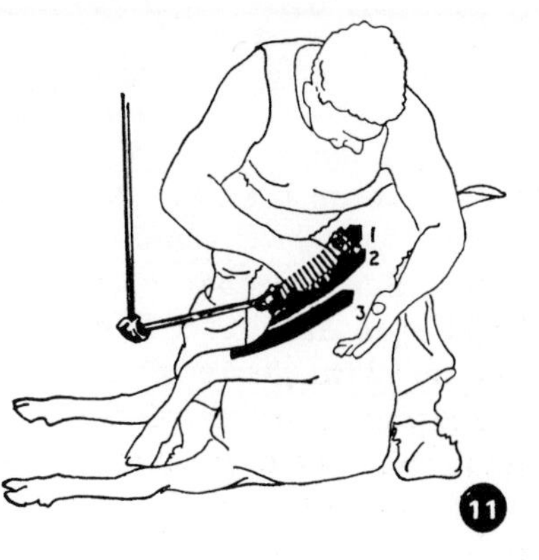

12. At last blow on shoulder, stretch leg up and turn in to flatten the shoulder joint thus allowing an easy blow.

13. Left foot under shoulder of the sheep, right knee to come down to press lightly on sheep's belly, left hand to move to grip behind sheep's ears.

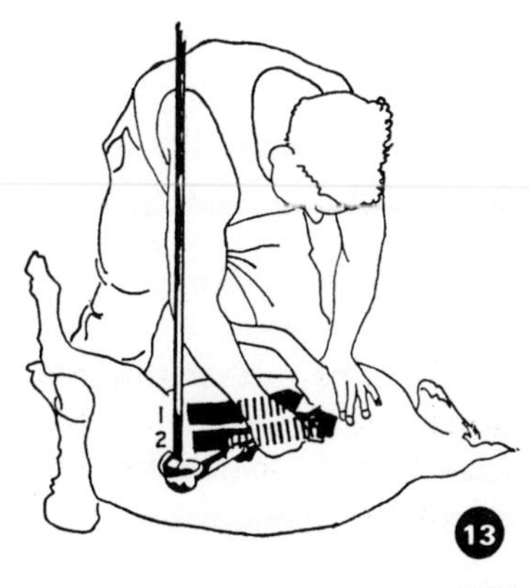

14. Long blow starts with full comb on the body, finishing with half comb on neck, thus keeping blows level with backbone. With second blow, shearer's right leg is placed over sheep's two hind legs thereby allowing the sheep to roll round.

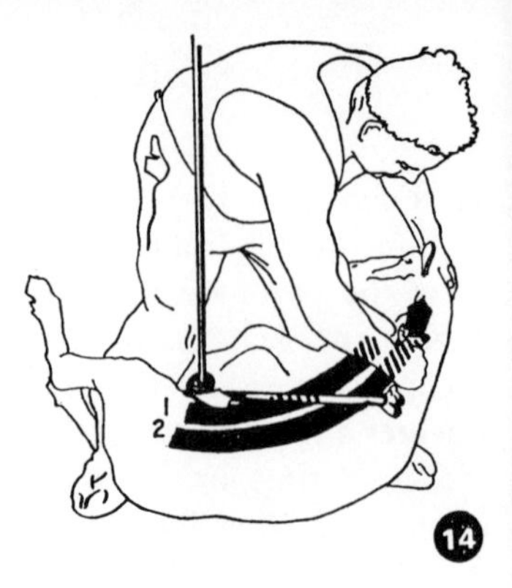

15. Shearer's left leg to be on an angle (and not straight up and down), with left foot under sheep's shoulder.

16. The sheep is rolled around on the shearer's leg, with its brisket turning to the floor. When coming off the long blow, shearer's left leg turns sheep further towards power shaft which allows right knee to come into back of sheep's head.

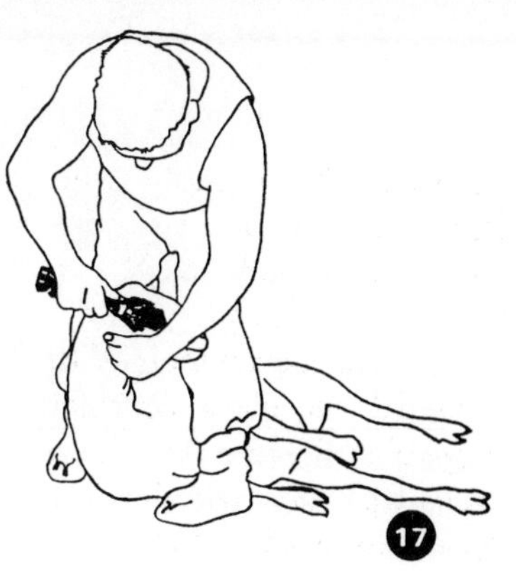

17. The last cheek . . .

18. . . . shearer's left hand pushes on the poll of the head.

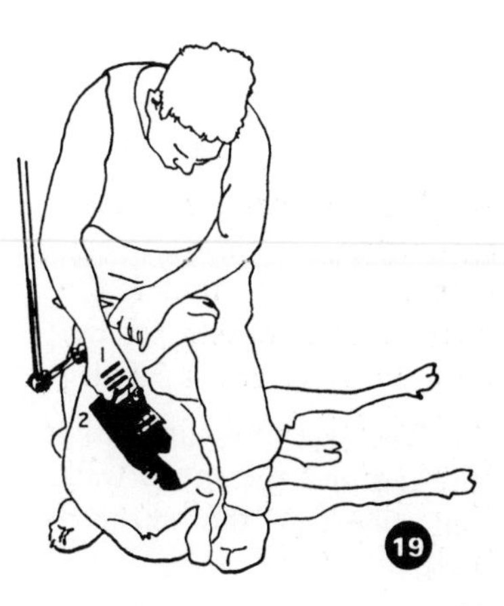

19. Blows then run down finishing inside the brisket and out the shoulder inside the front leg.

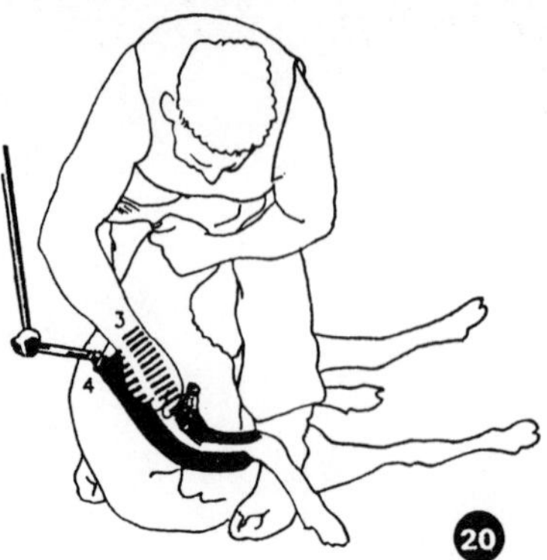

20. When the shoulder is cleaned, shearer's left leg steps out behind the sheep, its head being held between shearer's knees.

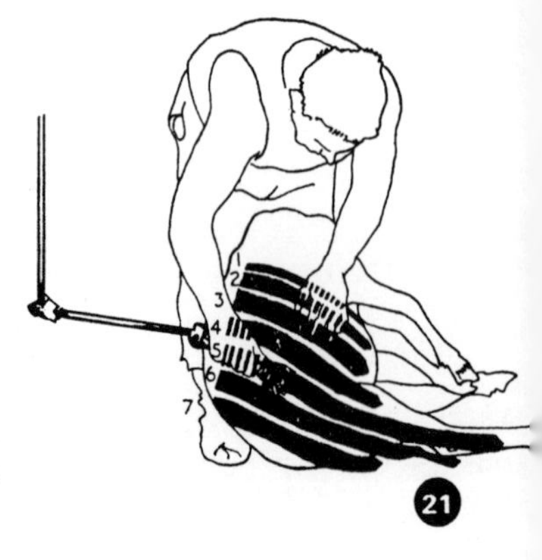

21. From the shoulder, angle blows in downward direction. Sheep is slightly bowed out by slight pressure of the knees.

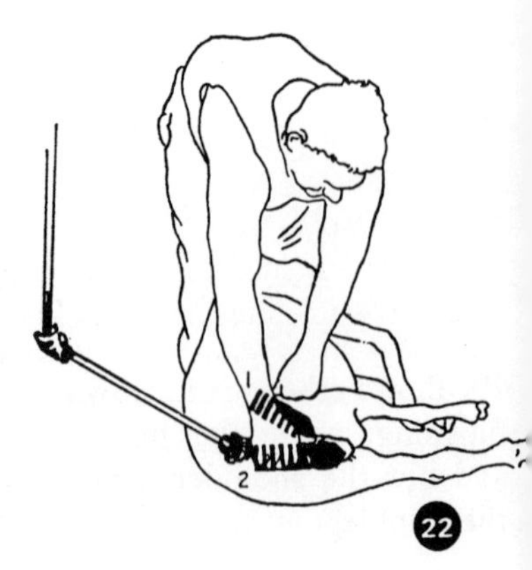

22. The last blow finishes at the tail and the sheep walks out between the shearer's legs.

Thirteen Wool

Wool is the most valuable product produced on the farm in the sense that no other farm commodity sells for as high a price per kilogram. It is important, therefore, that the producer takes great care in handling and presenting his clip. The first part of this chapter explains the more widely used wool terms, and briefly mentions the manufacturing process. The latter half deals with the care and presentation of wool on the farm.

Wool terms

Bradford spinning count from 'top' wool or 'counts'

During the manufacturing process wool is combed out into a thick ribbon and then wound into a ball, which is called 'top' wool. The top wool is then spun into hanks, each hank being 510 m long; the more hanks produced per 450 g of top, the finer will be the wool. The count of 60s means that each 450 g of wool yields sixty hanks, or sixty times 510 m.

In Australia and New Zealand, many of the Merino sheep produce wools as fine as 80s top, while in the British Isles breeds like the Scottish Blackface produce coarse wool which grades at 28s–30s.

Fine wools are more valuable than coarse wools; they are used for fine hosiery, babies' garments, vests and fine cardigans. Coarse wools are less valuable but are hard-wearing, and used for overcoats, blankets and carpets.

Length and strength

If we examine a piece of wool under a microscope we can see

that it consists of thousands of individual fibres, all interlocking with one another. On further examination we shall see that

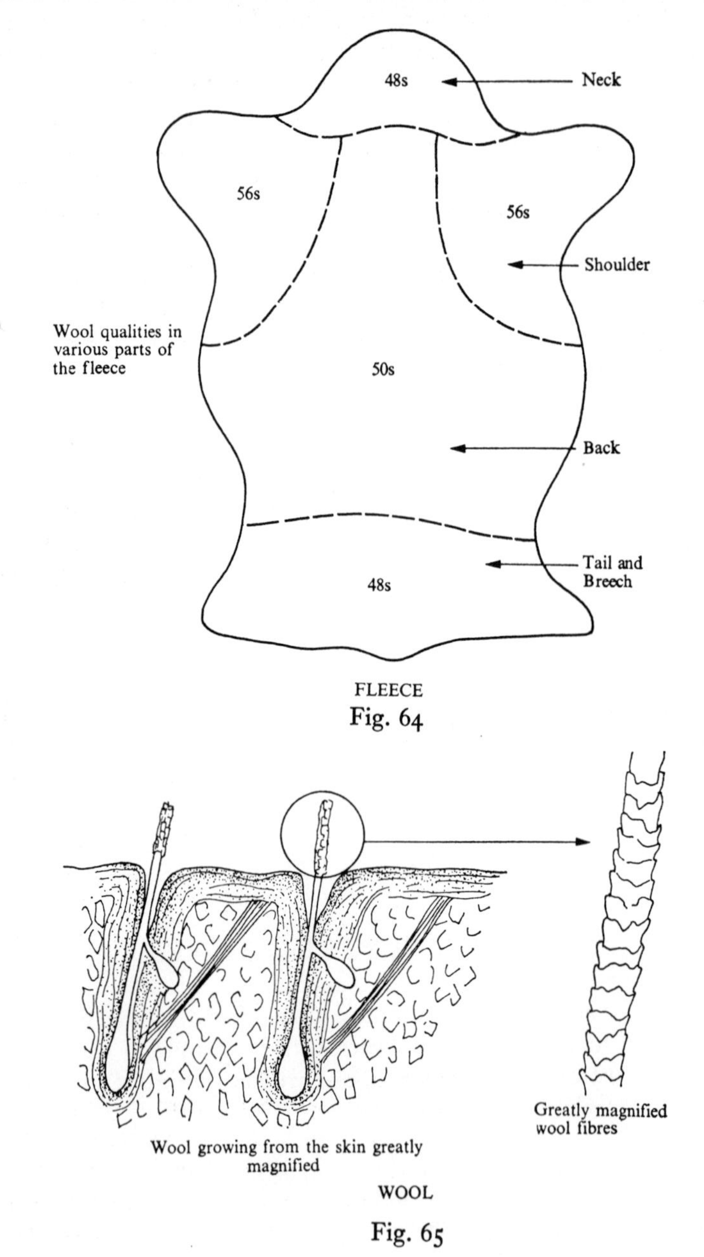

FLEECE
Fig. 64

Wool growing from the skin greatly magnified

WOOL

Fig. 65

each fibre consists of 'scales' or cells joined together. Fine wools have more scales than coarse wools.

It is these scales growing from the sheep's skin which gives the fleece its length and strength. Anything which interferes with this growth causes a weakening of the fibre, and, in severe conditions, growth may cease altogether, causing the fleece to break. For example, ewes that have a difficult lambing may later lose their fleece owing to the check of body condition and general health. Sheep that are healthy and fit produce the strongest fleeces.

The length of wool measured from the skin to the outward tip is called the staple length. The staple length of British sheep varies from 50 mm–76 mm for short wools up to 300 mm–500 mm in the case of longwool breeds.

Elasticity

One of the outstanding characteristics of wool is its elasticity. A damp woollen garment may be stretched up to 170 per cent of its length, and will then return to its normal shape when dried and pressed. Elasticity is detected by feeling the springiness of the wool.

Warmth-retaining qualities

Wool is an ideal insulator of body heat. It keeps sheep protected from rain and cold, and many people wear a woollen vest in winter or a woollen pullover or cardigan to retain the heat manufactured by the body. Wool will absorb perspiration or light showers of rain without the moisture being detected by the human touch. In fact wool can absorb up to 30 per cent of its own weight of moisture without being detected. It is for this reason that you should always 'air' clothes well before you wear them.

Colour

Pure white wool is by far the most valuable, and preferred by most manufacturers. White wool which is lustrous (bright) is preferred to white wool which is dull. Bright white wool can be used in its natural colour, whilst a dull white must be dyed. Any grey, black or red kemp fibres found in a white fleece will detract from its value.

Crimp

Crimp is the natural wave or corrigation in the staple. Merino sheep have around 22–28 waves per 25 mm, whilst coarse wools may have only 8–10 crimps per 25 mm length. Crimp is not found in other animal fibres in the same pattern; in fact most animal fibres are straight.

Crimp is usually associated with fineness of fibre, so that the more crimps per inch, the finer the wool. Crimp is also associated with elasticity. In fact, we may generalise and say that if a fleece shows a good distinct, even crimp, then it indicates a well-grown, sound fibre of uniform diameter.

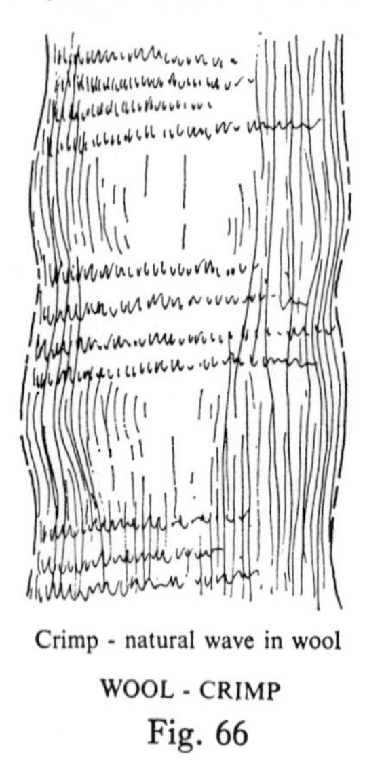

Crimp - natural wave in wool

WOOL - CRIMP

Fig. 66

Softness

All wool is soft to handle when compared with other fibres, but there is a tremendous variation in the softness of fleeces between breeds, and even between individual sheep within a breed.

Down breeds like the Southdown produce much softer wool than the mountain breeds, which grow strong but 'harsh' or 'wiry' wool. However, in breeding our mountain sheep, we must always remember that harsh wool will shed off the rain better than a soft wool. Softness is associated with crimp and fineness of fibre. Hence, the Ryeland and Southdown produce our best wools with counts of up to 60s and crimps of 18–20 per 25 mm length.

Summary

To conclude our discussion on wool terms we will look at wool through the wool grader's eyes. The most important factor which he looks for is fineness of fibre, since the finer the wool, the higher will be the Bradford spinning count. Next he will look for crimp which he associates with strength and elasticity. This is important because the wool must stand up to the machinery used in cloth making without breaking unnecessarily. Length of staple will depend upon the breed: Shortwool breeds produce 50 mm–76 mm staple, whilst Longwools range from 300 mm–500 mm. Colour should be uniform, a bright white (lustre) being preferred. Discolourations from dirt, grease, paint or string must be avoided. The wool should also be free from 'kemp', black or grey fibres, or any scabs from a skin disease.

Wool may be clipped in its natural state, or washed before shearing. Unwashed wool is referred to as 'greasy' wool.

Wool on the farm

We can improve the quality and presentation of wool on the farm in three ways; firstly by good shearing, secondly by taking care in handling and packing, and thirdly by selection of breeding stock.

Washing

The practice of 'washing' sheep before shearing has almost disappeared on lowland farms, although many upland farmers who have running streams on their farms still continue to wash them.

The secret of washing is a copious supply of running water, and where a farmer is able to dam a stream and has the necessary handling pens, washing is well worth while. Washed wool fetches a better price than greasy, dirty wool, and is much easier to shear. A further advantage is that the combs and cutters of the shearing machines will run more smoothly in washed wool than dirty.

Washing should take place about four days before shearing, as this will allow the fleece to dry thoroughly and the yolk to rise. If shearing is delayed too long after washing, the fleece becomes contaminated again, and the owner loses the washed wool premium. Most farmers, however, think that the loss of weight (dirt) and labour cost are not offset by the premium, and so they clip greasy wool.

Wrapping the fleece

With the exception of mountain breeds, all fleeces should be wrapped with the 'skin' outside. This is achieved by laying the fleece out on a clean floor or table with the skin side downmost. Remove all pieces of straw, brambles, etc., and then turn the

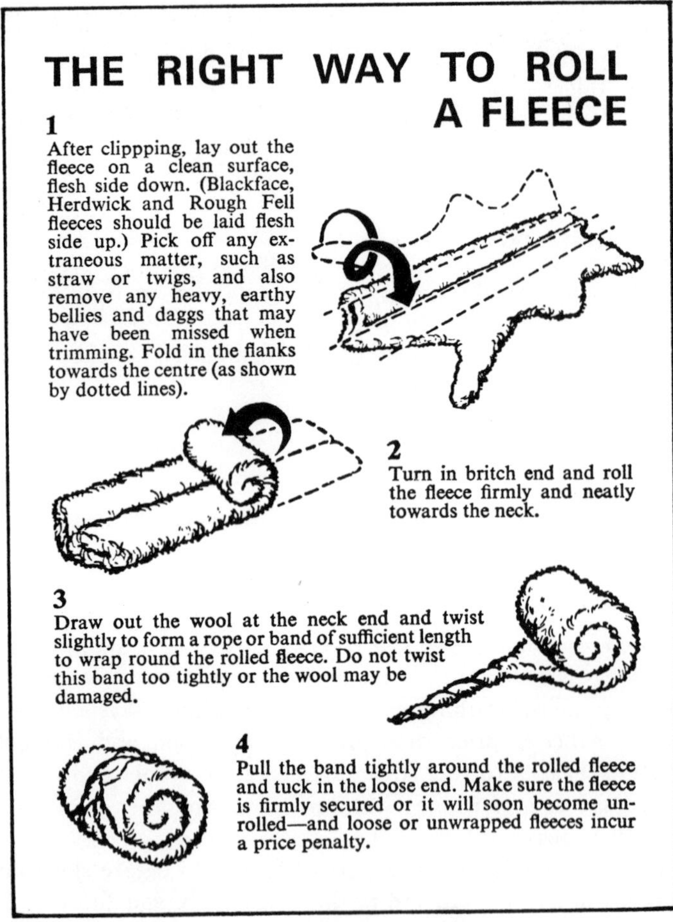

Fig. 67

sides inwards. Starting at the breech end, roll the fleece tightly until you reach the neck; then twist the neck wool into a 'rope' and bind it round the fleece. A well-rolled fleece is easy to handle by the grader, and if wrapped as described will show the shoulder wool outmost, which of course is the most valuable part of the fleece.

Rolled fleeces should be packed in a woolsack. Daggins, clags and 'odd bits' should be packed separately.

Summary

Avoid using straw for 'bedding' your penned sheep before shearing. Either pen them on slats, or use wilted nettles.

Never tie up a fleece with binder twine (hemp).

Always use an approved washable paint for brand-marking your sheep. Avoid pitch or oil based paints.

Dag your flock well in advance of shearing.

Pack the fleeces in wool sacks and tie a label with your name and address inside the sack in case your wool goes astray.

Store the wool sacks on a dry floor, preferably with a sheet of polythene underneath to stop rising damp.

Fourteen Handling Pens and Equipment

Where a flock comprises fifty or more ewes some form of sheep-handling equipment should be provided. Smaller flocks scarcely justify the cost of handling pens, and existing cattle pens are usually adequate. Handling pens may be either permanent or portable, the latter being extremely popular, since they may be transported easily around the farm.

PORTABLE SHEEP HANDLING EQUIPMENT
Fig. 68

Permanent handling pens

These may be built on any accessible site, where concrete floors can be laid and adequate drainage facilities made. The pens will consist basically of a collecting pen, for gathering the flock; a drafting race for sorting, which may also be used as a foot-rot bath; at least two sorting pens; and a dipbath or spray race.

Fig. 69 shows a very simple layout.

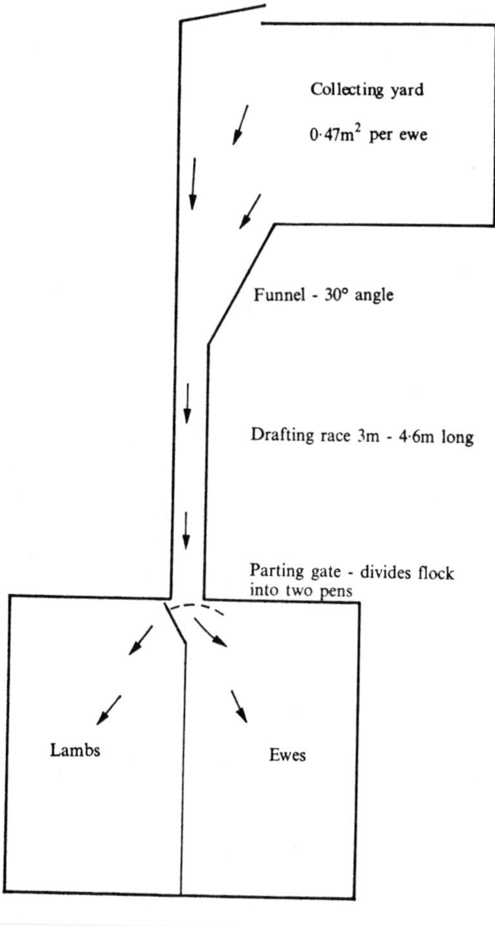

SIMPLE LAYOUT

Fig. 69

Collecting yard and holding pens for adult sheep

Sheep require 0·4 m²–0·46 m² per head of standing room. This means 100 ewes will need 40 m²–46 m², or a pen 6 m × 7 m. The sides may be constructed of post-and-rail, weld-mesh, corrugated sheets, or sheep netting. Post-and-rail is recommended because it is relatively inexpensive and easily erected by the farm handyman.

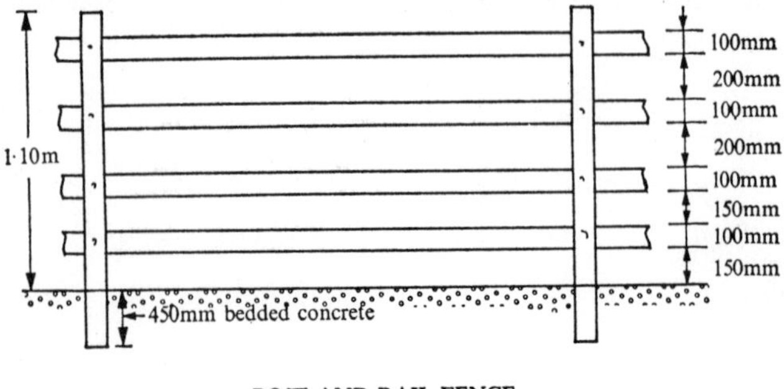

POST AND RAIL FENCE
using 100mm x 25mm rails on 150mm x 75mm posts

Fig. 70

The posts should be at least 100 mm × 75 mm and the rails 75 mm–100 mm × 25 mm. Rough-sawn oak posts and larch or similar softwood rails are most satisfactory.

The posts should stand 0·9 m–1·0 m above the ground, depending upon the breed of sheep. Downbreeds are docile and less likely to jump fences than mountain breeds.

Four rails should be used, suitable spacing being 150 mm between the bottom two rails and 200 mm between the top two rails. The narrower gap at the bottom prevents baby lambs from straying.

Floors

Ideally, the floors should be made with concrete, and sloped well for drainage. 150 mm of rammed hardcore covered with 50 mm–75 mm of concrete is sufficient for sheep. As concrete is expensive, many farmers use 50 mm–75 mm of washed gravel over the rammed hardcore, or 150 mm of broken limestone. The latter is claimed to assist in controlling foot rot. The washed gravel will keep clean if the pens are in the open, for, when it rains, the sheep's droppings dissolve and drain through the pebbles.

Drafting race

This is used for sorting the flock and as a holding pen for tasks like dosing, injecting or checking ear marks. The race is really a funnel through which the sheep pass in single file. By operating a swinging gate, the shepherd is able to direct the sheep into separate pens.

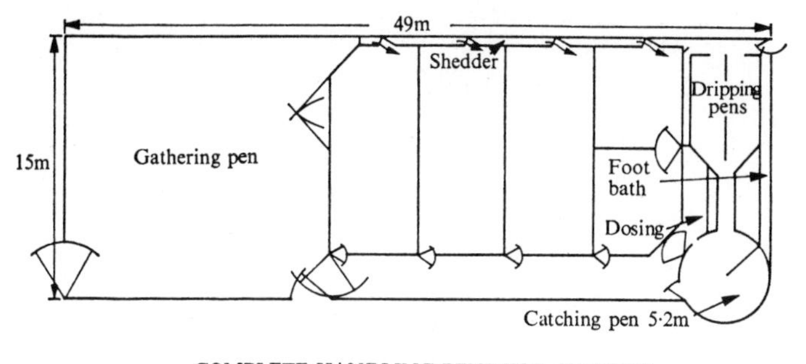

COMPLETE HANDLING PENS FOR 400 EWES

Fig. 71

The race should be approached by a Y-shaped crush that is not more than 1·8 m wide. This width will allow a man to drive the sheep single-handed into the race. The sides of the crush should be angled at 30 degrees to the race. This will allow the sheep to enter the race singly. Wider angles than 30 degrees tend to invite two sheep to enter the race together, or the sheep will turn round at the race entrance and walk back down the crush.

The sides of the race and funnel should be solid to prevent the sheep looking from side to side. The normal width for a race is 450 mm for mountain breeds and 500 mm–550 mm for Down breeds. The minimum length is 3 m but it is much better to build your race 6 m–8 m long. This will speed up drafting and allow more sheep to get into the race for routine tasks like dosing and injecting. It is well worth while fitting guillotine gates at each end of the race, which can be operated by pulling a nylon cord. This will greatly help the shepherd when working single-handed.

Foot bath

This may be built separately or included in the drafting race. However, since sheep are reluctant to walk through water, it is better to have a separate race and foot bath.

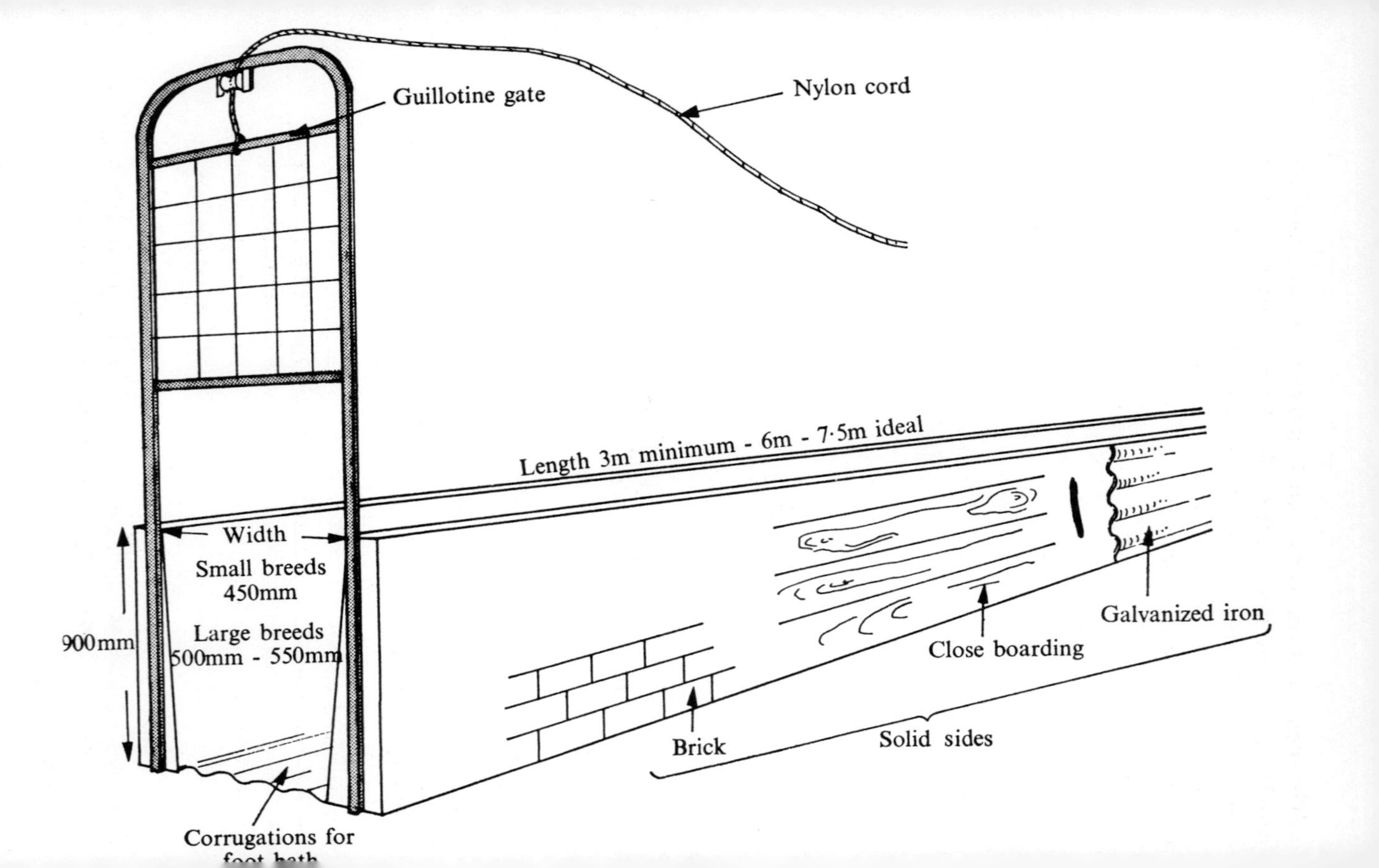
Guillotine gate
Nylon cord
Width
Small breeds
450mm
Large breeds
500mm - 550mm
900mm
Length 3m minimum - 6m - 7·5m ideal
Galvanized iron
Close boarding
Brick
Solid sides
Corrugations for

The floor may be constructed of concrete. Corrugations should be moulded into the floor, for these will open the clays of the foot and allow better penetration of the antiseptic. Corrugations can easily be moulded by simply laying a piece of corrugated asbestos sheeting over the wet concrete when it is laid.

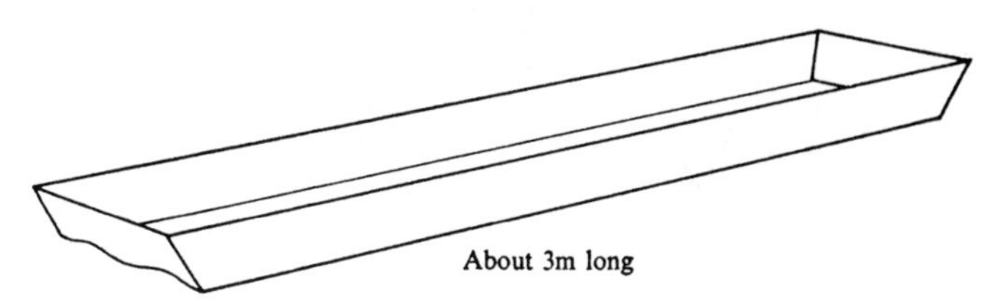

PORTABLE FOOT BATH

Fig. 73

Alternatively, a galvanised bath may be used. Suitable measurements for race and bath:

Length	3 m–6 m depending on size of flock
Width	450 mm–500 mm at top depending on breed
	300 mm–350 mm at base
Height	800 mm–950 mm

Dipping bath

There are three types of dipping baths available to the farmer:

1 Short swim bath
2 Long swim bath
3 Circular bath

The short swim bath is the most widely used. It is suitable for flocks of up to 1000 sheep. The bath may be purchased as a complete unit, made with galvanised iron, in various sizes from 473 litres to 1820 litres; or, alternatively, you can build your own dip, using concrete and brick.

The long swim bath is deeper as well as longer than the short swim bath. The extra depth allows an automatic tipping device to be used. Long swim baths are only economic for flocks of at least 1000 sheep.

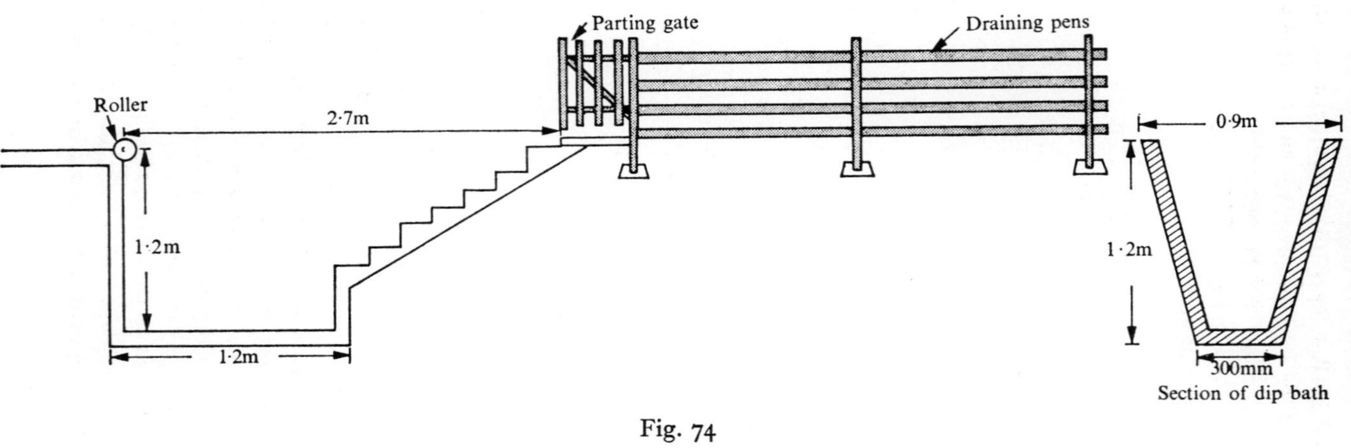
Roller
2·7m
Parting gate
Draining pens
1·2m
1·2m
0·9m
1·2m
300mm
Section of dip bath

Fig. 74

The circular bath is used with large flocks of 1000 head or more. The shepherd stands on an island in the centre of the bath and controls the sheep as they swim round.

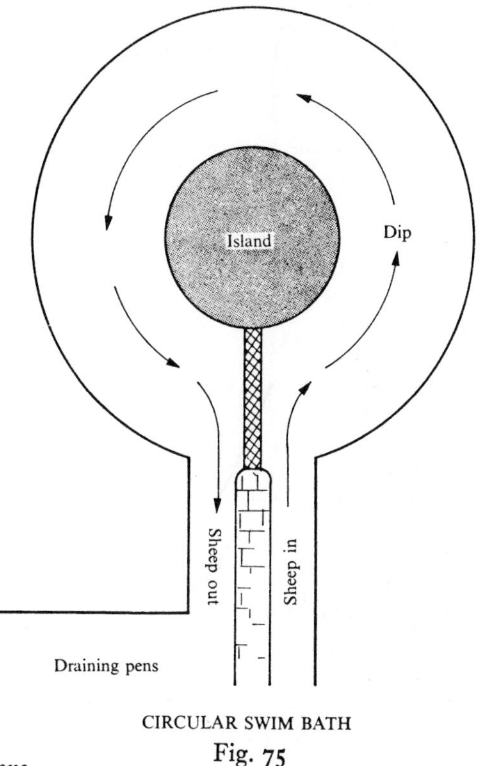

CIRCULAR SWIM BATH

Fig. 75

Draining pens

When a sheep leaves the bath it is carrying about 22 litres of dipping fluid. The sheep will soon shake itself and will lose about 18–20 litres of the fluid in a very short time, only 2–4 litres being retained in the fleece. The purpose of draining pens is to catch the surplus dip and return it to the sheep bath.

Two draining pens, side by side, should be erected at the exit from the dipping bath. Each pen should have a concrete floor, with a steep slope back to the bath. When dipping is in progress, each pen is used, holding the sheep that are draining. When the second pen is full, the first is emptied, and so on. In this way each sheep is allowed 5–10 minutes to drain before being returned to the field. Each draining pen should be large enough to hold about twenty sheep.

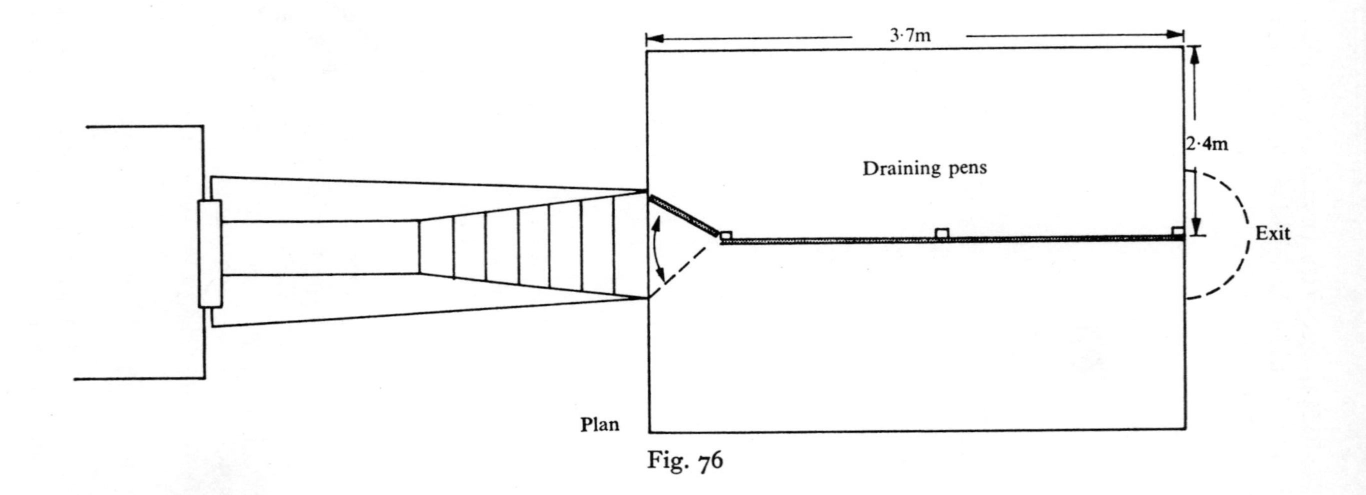

Fig. 76

Fifteen Sheep Housing

Winter housing

Lamb wintering sheds have been used in the uplands areas for some time, but recently there has been considerable interest in housing the breeding flock. Some flock masters house their ewes from December until the spring whilst others only bring the ewes inside for lambing. The buildings need not be expensive, unused cattle yards, pole barns, and recently polythene houses have been used. The important thing is that the house is dry, airy and offers protection against the rain and cold winds.

There are numerous advantages in housing breeding sheep: the land is not poached during prolonged wet weather, more control over the sheep is possible especially at lambing time, and mortality in young lambs is reduced considerably.

The disadvantages are the cost of buildings if new ones have to be erected and the risk of disease, especially foot rot if a slatted floor is not used. Great attention must be paid to the ventilation of the building in order to prevent air-borne diseases, and making the flock uncomfortable.

The site

When erecting a new building one should look for a level free-draining plot that has access to the fodder store, roads, water supply, and if possible some small sheltered fields where ewes and lambs may be turned for a few days to 'acclimatise' when they are first turned out in the spring.

Materials and construction

Wood, concrete blocks, straw bales, polythene, oil-tempered hardboard, or corrugated iron may be used for cladding, the

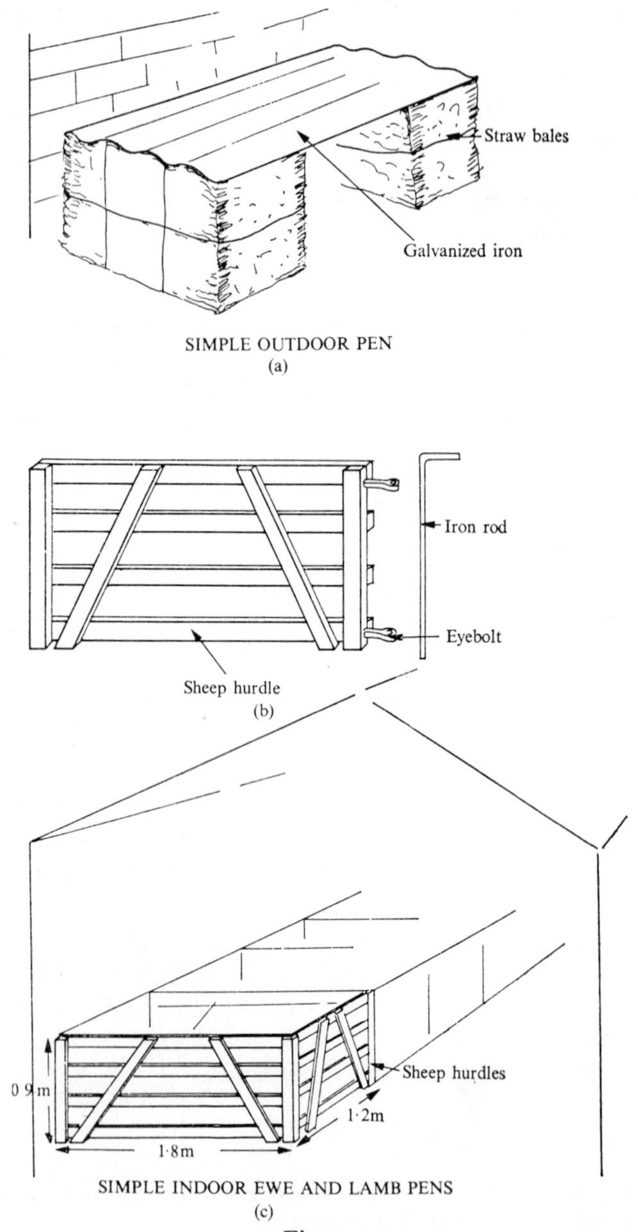

SIMPLE OUTDOOR PEN
(a)

(b)

SIMPLE INDOOR EWE AND LAMB PENS
(c)

Fig. 77

main aim being the cheapest material available. Pole barns with corrugated-iron roofs and straw bale sides 2 m high make excellent houses. Each ewe will require about 1·5 m^2 of floor area if straw is used for bedding. With slatted floor houses the floor area may be reduced to around 1 m^2 per ewe. Slatted floors are made with wooden battens measuring 60 mm × 50 mm spaced 15 mm apart. The battens are nailed on to 100 mm × 50 mm cross timbers and should stand on 500 mm pillars to allow for a dung pit.

Troughs and hay-racks

Adequate trough space is vital in order that timid sheep may obtain sufficient food. Allow 304–380 mm for store lambs and 450 mm for ewes. Hay-racks may be made with wire sheep netting shaped into a letter U and suspended from the roof. Alternatively a combined hay-rack and trough may be used. Always remember to ensure that a clean unfailing water supply is available.

In wintering ewes

In winter ewes are normally housed in late December or early January, depending on weather conditions, and the available grass supply.

Before the flock is housed it is important that they receive good quality hay or silage to get them accustomed to eating dry food.

Particular attention should be paid to their feet. Any sign of foot rot should be treated immediately and the entire flock should be walked slowly through a foot bath containing 6–10 per cent Formalin before they are housed.

Also check the water supply to see that it is protected against frost and make sure that water troughs and bowls are thoroughly cleansed, because sheep will not drink stale water and are reluctant to use dirty containers.

Once housed the flock will quickly adapt to their new surroundings and as pregnancy advances will become noticeably docile. Walk quietly amongst the ewes and they will soon come to recognise you. This will later be invaluable, for when lambing starts you will be able to walk through the pens with the minimum of disturbance to the ewes.

Keep a watchful eye on the sheep's dung, especially if you are feeding hay only. Sheep fed on silage or hay and roots, produce large quantities of 'wet' faeces, whilst those receiving fibrous hay or chopped straw tend to become constipated or produce a small quantity of 'dry' faeces and little urine.

To avoid constipation, ewes may be offered some 'wet' sugar beet pulp. About 125 g soaked in water for twelve hours is an ideal feed. It acts as a succulent food and also adds valuable energy to the diet.

Should you wish to dose the flock against roundworms, this task is best undertaken two to three weeks after housing. In this way you will control all the eggs which were picked up just prior to housing and then there will be a good 'kill' of all worms present. Providing, you use plenty of bedding there will be little risk of the ewes becoming re-infested with worms whilst they are housed.

Feeding

Hay is usually offered twice daily in either Norwegian Hay Boxes or wire mesh hay-racks. Only the best quality hay should be offered, and dusty and mouldy hay avoided. Mouldy hay can cause ewes to abort.

Once 'steaming up' commences it is advisable to divide the ewes up into groups of 20 to 30 per pen. This will help to prevent bullying and ewes heavy in-lamb, getting knocked about whilst coming to the troughs at feed time. Also when the lambing commences there will be less risk of mis-mothering if there is only a small number of ewes in each pen.

Lambing

Ewes will usually find a corner in the pen in which to lamb. They should be left undisturbed until the lambs are born, and then quietly and gently removed from the pen and taken to a nearby small individual pen, and kept there for the next day or so.

Newly born lambs are vulnerable to many bacterial diseases, particularly 'joint-ill', and should always be treated by spraying or dipping the navel cord in iodine or a suitable antiseptic. The ewe's afterbirth or cleansing should be removed to prevent the build up of disease in the building and individual pens should be cleaned out and disinfected between successive lambings.

Plenty of clean bedding should be used daily in the main pens throughout the lambing period.

After lambing

Generally speaking farmers prefer to turn the couples out to grass when the lambs are one or two days old. This will keep them fit and lessen the risk of disease. However, in severe weather the couples may be kept indoors, but must of course be given plenty of room to avoid mis-mothering and lambs being crushed.

Conclusions

Advantages of housing

Better working conditions
Avoidance of poaching and overgrazing—early grass
Intensification where winter grazing determines stocking rate
May permit winter cultivations on old pasture before frost

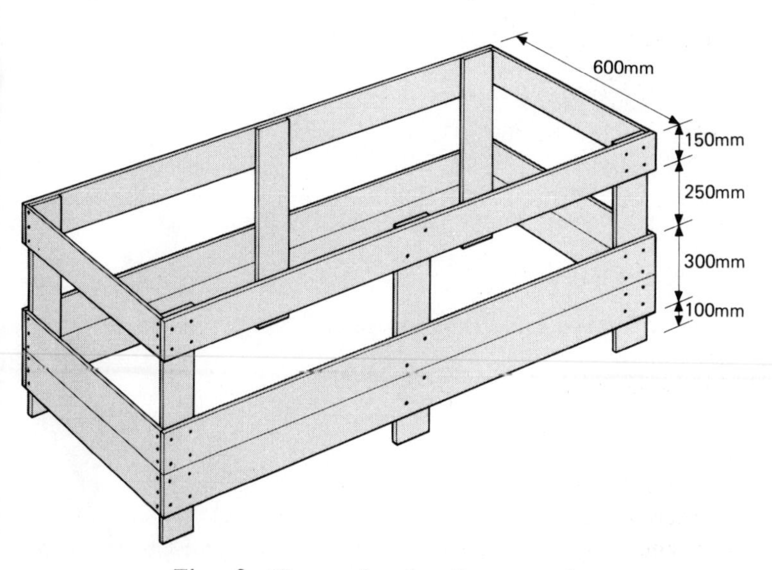

Fig. 78 **Norwegian hay-box trough**

Should reduce lamb casualties through better shepherding
May increase ewes working life—lower depreciation
Better utilisation and supervision of expert labour
Better feed utilisation
May permit out of season lambing
May be able to pen and feed by condition score
Flock inspection easier
May be cheaper to house hoggets than to away winter
May offer alternative of fat to store market for wedder lambs
Ewe mortality should be reduced, especially hill ewe

Disadvantages of housing

High capital involvement if new buildings erected
Extra maintenance and possibly building charges
May increase disease problems. Therefore high preventative medicine costs.
In mild winters feeding costs may be higher
Higher conserved feed requirements
Higher lamb mortality if hygienic standards are low
Wool loss may be marked and fleece depreciated

Relevant factors to consider

Housing costs—including bedding
Winter feed costs
Ewe and lamb performance
Influence of stocking density
Labour costs and working conditions
Other economic uses for building to spread costs
Alternative use of winter accommodation land

Alternative uses for building

Covered handling and shearing pens
Implement storage
Turkey rearing
Calf rearing
Short-term grain storage
Fodder storage
Barn hay drying
Two throughputs of sheep a year with early and late lambing season

Sixteen Common Ailments and Diseases Affecting Sheep

Foot rot in sheep

It is often said that a shepherd's efficiency may be judged to some extent by the amount of foot rot in his flock. Foot rot is one of the most widespread diseases affecting sheep, and is found in all types of flock, from mountain to lowland, and in all ages of sheep, from old rams to young lambs. It may be in an acute form in some sheep, and chronic in others. However, if the flock master has the knowledge, the new medicaments, and above all the determination to stamp out the scourge, he can eventually eradicate the disease from his flock.

Cause

The disease 'foot rot' is caused by a bacterium called Fuziformis nodosus. The bacteria may live in an 'inactive' state in cracks and crevices of a sheep's hoof for up to nine months. If it is removed from the hoof and passes on to the ground, then it cannot live for more than seven days. These bacteria only attack sheep and goats, and no other animal.

Conditions of occurrence

Sheep of all ages are susceptible, but prolonged wet weather, badly drained land, and rank long wet grass all tend to encourage the disease. It should be remembered, though, that the disease will not occur if the bacteria are not present.

Symptoms

The first sign of foot rot is lameness in *several* sheep in a flock.

(Individual lameness may arise from other forms of injury.) Two or more feet may be affected, and in some cases the sheep are only able to graze by kneeling on their elbows. Untreated sheep will rapidly lose bodyweight and condition.

Diagnosis

Infected sheep should be caught and turned up for inspection. Remove any mud that may be lodged between the clays and then carefully pare away any overgrown nail. There is a characteristic fishy smell which will quickly be detected in infected feet. Any pus (evil-smelling fluid) must be cleaned out and the wound painted over with a mild antiseptic cream or sprayed with either an antiseptic or antibiotic (Chloromycetin) spray: e.g.

2% Formalin = 3 teaspoonfuls to 0·5 litre water
10% Dettol = 4 tablespoonfuls to 0·5 litre water

In severe cases the foot may be bandaged, or covered with a specially made rubber boot. In such cases the covering should be removed and fresh dressing applied every 3–4 days. Under no circumstances should strong caustic chemicals be used, for such substances will destroy healthy tissue and hinder rather than encourage healing.

Prevention and treatment

The entire flock should be passed slowly through a foot-rot bath once a month as a precaution, and once a week if an outbreak occurs. The bath should contain either:

10% copper sulphate = 0·45 kg copper sulphate to 4·5 litres of water

or 6% Formalin solution = 0·25 litre of commercial Formalin to 4·5 litres of water.

Afterwards the sheep should stand on clean, hard concrete for about an hour, and then be turned on to a clean, rested pasture, i.e. a field that has carried no sheep for at least fourteen days.

Note

1 The shepherd should frequently wash his hands and sterilise his knives and secateurs after dealing with each infected sheep.

2 Do not allow the flock to walk over infected soil, or through infected pens and gateways when returning them to pasture.
3 Change the flock on to a clean pasture every seven days, and do not return them to the field for a further fourteen days, in which time most of the bacteria will perish.
4 Finally, it must be stressed that foot rot can, and must, be controlled. It calls for perseverance on the part of the shepherd. He must constantly be on the look out for infected animals, and take immediate action once the disease is diagnosed.

Diseases caused by Clostridial bacteria

Clostridial diseases are a most important group of diseases. The Clostridial bacteria are anaerobic, which means that they live without free oxygen. Disease-producing anaerobes are able to produce spores which can exist in the soil for many years, and then quite suddenly become active and affect animals. Once a spore becomes active, it rapidly changes into a bacterium and produces toxins. The toxins are the poisonous substances that kill the affected animal.

We can see then that it is quite possible for spore-forming bacteria to live on a farm, either in the soil or beneath dirt or dung, or in cracks and crevices of old buildings for a very long time, and then quite suddenly to flare up and cause severe losses amongst stock.

The Clostridial diseases that affect sheep are lamb dysentery, pulpy kidney, tetanus, enterotoxaemia, black quarter and struck. Fortunately, we have excellent preventatives for these killer diseases in the form of vaccines and anti-sera.

Pulpy kidney caused by Clostridium welchii type D

In most areas where sheep are kept intensively, or run on improved pastures, new leys, etc., pulpy kidney disease is found. The disease affects vigorous, strong, growing lambs, usually the best ones in the flock. Affected lambs develop stiffness in the limbs very similar to lock-jaw, they may froth about the mouth, or in acute cases are just found dead. Losses are most frequent in the spring.

Diagnosis

Open the dead lamb and inspect the kidneys; if they are found to be pulpy and jelly-like, you can be fairly sure the lamb died with pulpy kidney disease.

Cause

It appears that more than one organism may be involved, and it may be *Clostridium welchii* type D in the bowel, for this has been shown to be responsible in this country. In all probability the high protein in the spring grass acts as a predisposing factor, rendering the condition of the bowel particularly suitable for the bacteria to infect.

Pulpy kidney can also attack adult sheep, when it is referred to as enterotoxaemia.

Prevention

Where the disease is known to be on the farm, follow one of the following three methods:

1 Vaccinate the yearling ewes with 5 c.c. *Clostridium welchii* type D vaccine just before tupping, a further 2 c.c. within a month, and a 2 c.c. booster just before lambing. Ewes in their second year only need the 2 c.c. vaccine 10–14 days before lambing. This method will prevent the disease for the first 12–16 weeks after birth.

2 Vaccinate all lambs soon after birth with 2 c.c. per lamb.

3 Use combined sheep vaccine. These now give protection against lamb dysentery, pulpy kidney, tetanus, struck, blackleg and braxy. If the ewes are vaccinated according to the recommended dose, protection will be given to the ewes and their offspring until they are four months old.

Treatment/prevention

A sudden outbreak in a non-protected flock can be checked by moving all the sheep on to a poorer pasture, and then injecting them with pulpy kidney anti-serum. The serum is immediately effective, and will last for 2–3 weeks.

Lamb dysentery—Clostridium welchii type B

This is a highly infectious disease that is found widely in the lowlands and particularly in the hill sheep areas. The disease usually attacks lambs under one week old, generally on the second or third day. Affected lambs soon become dull, refuse to suck, and may bleat with pain. There is usually diarrhoea which is a greyish colour and may contain blood. Within a short period the lamb is unable to rise and death soon follows.

Prevention

Routine vaccination of breeding ewes is now practised by most farmers. Should by chance a flock have not been vaccinated then the lambs may be injected with lamb dysentery serum within a few hours of birth.

Enterotoxaemia—Clostridium welchii type D

This disease of older animals is sporadic (occurring casually here and there) and found throughout Great Britain. It is usually associated with yearlings, but may affect adult sheep up to two years of age.

Incidence

Losses generally occur when the flock is changed from poor to richer herbage. The mortality rate is generally fairly *low* and variable. The first indication of disease is generally finding a dead sheep.

Prevention

Pulpy kidney vaccine for *Clostridium welchii* type D.

Enterotoxaemia may also be caused by *Clostridium welchii* type C. This is only found in certain parts of the country, notably in Romney Marsh in Kent, and in parts of North Wales. The disease is locally known as 'struck'. The prevention is to vaccinate with *Clostridium welchii* type C vaccine.

Black disease—Clostridium oedematiens

This affects adult sheep and is caused by *Clostridium oedematiens*. Infection is associated with an invasion of the liver by migrating, immature flukes which produce localised areas of dead tissue in the sheep's body in which the spores can germinate.

Incidence

Only found in restricted areas in the United Kingdom. It is dependent on migratory fluke larvae for its occurrence.

Symptoms

Sheep are frequently found dead. If found alive, sheep are disinclined to move, and thereafter pass into a recumbent position, sitting on their sternum, with outstretched forelegs. Death generally occurs without struggling.

Prevention

One approach to prevention is the control of the liver fluke. Active immunity may be obtained by the use of black disease vaccine in August, and again four weeks later.

Tetanus—Clostridium tetani

This disease affects a wide variety of animals, including humans. It is caused by *Clostridium tetani*. The spores infect a wound and multiply, producing a toxin.

Symptoms

Tetanus sometimes follows docking and castration in lambs. As in other animals, it can similarly follow surface wounds. The effect of the toxin is to produce a characteristic muscular stiffness which progresses to end in muscle spasms. In acute cases, animals go down, or stand with their legs outstretched. The muscles of the jaw may become rigid, and hence the name 'lock-jaw'.

Prevention

Commonsense precautions of cleanliness and hygiene at docking, castration and other operations can aid prevention. Vaccinate with combined vaccine.

Blackleg—Clostridium chavoei

This is a fatal disease of sheep, and is often referred to as post-parturient gas gangrene or wound gas gangrene.

Symptoms

The disease is generally associated with some form of external damage such as that following lambing, docking, castration, or the infliction of local wounds which become contaminated. It may also follow the use of a dirty hypodermic syringe.

The infected areas become either hot and swollen, or cold and very painful. Death occurs in 24–48 hours.

Post-parturient gangrene is found particularly in areas of Cornwall, Devon, Romney Marsh and the Welsh Borders.

It may be passed from ewe to ewe by the shepherd's hands if he has delivered an infected ewe.

Prevention

Use blackleg vaccine: but remember always to sterilise the hypodermic syringes, and to maintain cleanliness at castration, docking and other operations.

Braxy—Clostridium septicum

This is an acute disease of the digestive tract caused by *Clostridium septicum*. It is generally very rapid in onset, and the first symptoms are usually death.

Symptoms

The disease is caused by the invasion of the walls of the abomasum or fourth stomach by *Clostridium septicum* organisms. The bacteria multiply rapidly and produce toxins which are absorbed through the stomach wall into the blood-stream. The disease is often associated with frosty weather.

Prevention

Vaccinate sheep with Braxy vaccine. Two injections are given, four weeks apart in August–September.

Pregnancy toxaemia

Pregnancy toxaemia or twin lamb disease is a very serious acute functional disorder which may affect in-lamb ewes during the last few weeks of their gestation.

Cause

The cause is still not properly understood, but the evidence suggests that it is due to a deficiency of carbohydrates, i.e. temporary starvation.

Carbohydrates are stored in the body in the form of glycogen and are usually found in the liver and muscle. In response to the needs of the body, the glycogen is converted into glucose or blood sugar, which is readily assimilated by the body. The glycogen reserves at any one time are dependent upon the food intake of substances from which glycogen is produced, and upon the demands of the body for utilisation. During pregnancy, these demands are heavy because of the needs of the developing lamb. As pregnancy advances, the demand is increased. Glycogen stored in the muscle cannot be released as quickly as if stored in the liver, but plenty of exercise does speed up this process.

Occurrence

The disease is often associated with periods of severe climatic conditions—snow, severe frost, etc., or when sheep are excluded from their natural grazing. The disease only occurs during the last few weeks of pregnancy, and almost always attacks ewes with twins or triplets in their uterus.

Symptoms

The affected ewes appear dull and lifeless; they walk with an uncertain, sometimes staggering gait; they appear to be blind, and will often walk into fences or fall into ditches without appearing to see the obstacle.

The disease is frequently associated with the presence of acetone in the breath. As time advances, the ewe may be unable to rise, and death soon occurs unless the animal is treated or gives birth to her lambs.

Treatment

At the time of writing there is no really effective cure, although many treatments are being tried by the veterinary profession.

Warm glucose may be injected subcutaneously, and this has given reasonable results. It is best to seek veterinary advice.

Prevention

Regular exercise and feeding adequate amounts of a balanced concentrate diet, in addition to the normal winter grazings, is the best practice. The exercise helps to release glycogen from the muscle, and the concentrate ration will make good the extra demands made upon the ewe by the unborn lambs. Try, then, to walk the flock as much as possible during the last six weeks of pregnancy. This can be achieved to some extent by feeding hay in one field and concentrates in another. The ewes will wander to and fro.

Hypocalcaemia

Hypocalcaemia or lambing sickness is an acute disease which affects the nervous system of ewes just before or just after lambing. The disease is very similar to milk fever in dairy cows.

Symptoms

The ewe is found lying down and may be grunting. If not treated immediately, she will fall into a coma and later die.

Treatment

On veterinary advice you may inject calcium-boro-gluconate

subcutaneously, which should lead to quite a spectacular recovery.

Hypomagnasaemia

Hypomagnasaemia or 'grass staggers' is very similar to lambing sickness, although the attack is often acute, and the first symptoms are death. The disease may affect dry sheep, but is most common in lactating ewes during the early spring months. If the disease is detected early, then injections of magnesium boro-gluconate will be effective.

Prevention

Include magnesium-rich mineral mixtures in the concentrate rations fed for steaming up. With dry sheep, make mineral licks freely available.

Pine

Pine is a wasting disease which affects sheep in certain areas of the British Isles and in Australia. The disease is due to a deficiency of cobalt in the sheep's diet.

Symptoms

Pine is very similar to Johne's disease and a heavy infestation of roundworms. It usually affects young growing sheep, and may also affect cattle.

Lambs become dull and the eyelids appear pale; the fleece loses its lustre and later breaks. If left untreated, losses may be up to 30 per cent of the flock.

Treatment and prevention

Cobalt deficiency may be rectified by feeding cobalt in the concentrate diet or by dosing the flock with cobalt bullets. Cobalt sulphate may be applied to the pastures at 2 kg per hectare on farms where there is a high incidence of the disease.

Swayback

Swayback is a serious nervous disorder which may affect lambs soon after birth, and when about 3–6 weeks old. The disease is caused by a lack of copper in the diet.

Symptoms

The lambs have a marked lack of co-ordination between front and hind limbs and are extremely reluctant to try to walk. They stumble about only when roused by the shepherd. In acute cases, death occurs within a few hours, whilst in mild cases the lambs may live for several weeks.

Prevention

Specially formulated mineral licks containing copper sulphate may be fed to pregnant ewes.

Treatment

Seek veterinary advice.

Louping-ill

Louping-ill or 'trembling' is a serious disease caused by a virus that is transmitted from sheep to sheep by the sheep tick. The disease is widespread in Scotland and the northern counties of England. The disease affects cattle and pigs as well as sheep.

Symptoms

In the early stages of infection the sheep is characterised by a dullness and 'trembles'. There is a marked irregularity of gait, and the sheep may lose its balance. In the later stages paralysis of one or more limbs may develop.

Prevention

Sheep may be safely protected from louping-ill by vaccinating the entire adult flock with vaccine prepared from the virus. Lambs are not normally vaccinated until about twelve months old.

Treatment

There is no specific cure, but nursing the sheep indoors will sometimes lead to complete recovery.

Scrapie

Scrapie is caused by a virus which lodges in the brain and spinal cord, causing a serious nervous condition which eventually leads to death.

The virus may be transmitted from sheep to sheep when at pasture, in the act of mating, and from ewe to lamb. However, the virus has an incubation period of 18 months–2 years, so it does not affect lambs or yearlings.

Symptoms

The disease usually appears in odd isolated cases amongst the older sheep in the flock. The infected animal loses condition slowly at first, and develops an 'itch', the irritation causing the sheep to rub itself frequently against solid objects. In the later stages the sheep loses its balance and becomes excited. There may be scabs on the parts of the body that the sheep has rubbed, e.g. around the rump and parts of the head. The wool breaks and falls away.

The symptoms may show for as little as two weeks, but usually last from two to six months, terminating in death.

Prevention

There is no specific cure for the disease, but widespread research is taking place. One school of thought suggests breeding strains of sheep resistant to the disease.

Sheep scab

Sheep scab is a notifiable disease which can cause serious loss of body condition in sheep and in extreme cases death. The disease was eradicated in Great Britain for a number of years, but reappeared in 1973. Farmers who suspect that their stock are infected must notify the Police, or their Divisional Veterinary Officer.

Symptoms

In winter, affected sheep suffer intense irritation, mainly on the shoulders, back sides, and tail. They are frequently to be seen rubbing affected parts with the head or against posts, etc., or nibbling accessible areas with their teeth. When scratched they show signs of pleasure such as smacking the lips. In bad or untreated cases, scabs form and there is matting and loss of wool, sometimes over a wide area of the body. The irritation may interfere with normal grazing and resting to such an extent as to cause serious loss of condition, and in young lambs, even death.

Affected sheep may not show obvious signs of the disease in summer, when the mite lies dormant, but a close examination may reveal signs—such as thickened skin or small areas of scab formation—of the presence of mites and eggs in folds of skin around the eyes, within the ears, elsewhere on the head and in the groins.

Control and cure

The mite can be killed and the sheep cured by dipping in approved single-dipping type sheep dips. These contain an active ingredient which, after dipping, remains in the fleece more than long enough to kill mites which were in the egg form at the time of dipping and which emerge subsequently. Badly affected sheep should be hand-dressed to ensure the dip penetrates the scabs. The use of approved dips against the disease is required by law and it is important to ensure that the dip used bears a label stating that it is approved by the Minister of Agriculture, Fisheries and Food and by the Secretary of State for Scotland for use against sheep scab.

Dipping in an approved single-dipping type sheep dip will also protect unaffected sheep for several weeks and will normally assist in the control of other skin parasites such as lice and keds.

Dipping

It is virtually important that the correct dipping procedure should be followed. The exact capacity of the bath should be known. It should be emptied and a start made with clean water. Make sure you use an approved dip, follow the instructions on the label carefully and measure out the correct amount of dip. Stir the dip wash thoroughly before starting and after any break in the dipping. Sheep must be kept immersed in the bath for one minute. During this time the head must be ducked under at least once to ensure a thorough wetting of the head and ears and the sheep must be plunged or revolved to remove air pockets in the fleece so that the dip penetrates to the skin, where the mites live. The bath should be topped up regularly, again following the maker's instructions, and any manure and debris should be skimmed from the surface.

Contagious ecthyma (Orf)

Contagious ecthyma, usually referred to amongst shepherds as Orf, is a serious problem in lactating ewes and young lambs. The usual signs are blisters around the mouth of lambs and over the udder and teats of suckling ewes. Vesicules can also be found on the face, eyelids and feet of sheep at various times and in sheep of all ages.

Suckling lambs will refuse to eat and often premature weaning occurs. There may be some deaths following a rapid loss in body condition.

The disease is caused by a virus infection which leads to vesicules which quickly change to scabs that protrude from the skin. A hard brown mass of scab is easily recognised by the shepherd who should seek veterinary advice immediately.

Prevention and treatment

The usual prevention is for a veterinary surgeon to prescribe a 'live' vaccine which is applied to scarified skin, usually around the thigh area. Already infected animals are treated with broad spectrum antibiotic ointment or an antibiotic aerosol spray.

Bathing the infected areas with a mild solution of Dettol is also effective; but care is required on the part of the operator as Orf can affect humans.

Abortion in sheep

Many flocks experience the odd abortion but frequently abortion storms cause severe losses. Surveys suggest that the average abortion rate is between 1 to 10 per cent, while in individual flock outbreaks losses can be up to 40 per cent.

The term 'abortion' commonly means the premature birth of the lamb but losses also occur from failure of fertilisation of the ovum, death of the foetal lamb during the early stages of pregnancy resulting in reabsorption, or mummification and stillborn lambs. Finally, the birth of weakly lambs can result from adverse conditions of development of the lamb in the womb due to infection, or poor feeding.

Aborted lambs and their afterbirth should be promptly made available to your veterinary surgeon who will arrange to have a laboratory examination carried out. Most agents which cause abortion are to be found damaging either or both the foetus and placenta; hence in all cases a laboratory diagnosis is made by an examination of as many aborted lambs and placentas as can be submitted.

Common causes of abortion

Vibrionic abortion is widespread in distribution and is caused by a small bacterium which infects the placenta causing abortion during the last few weeks of pregnancy. Lambs can also be stillborn or born weakly and die.

In an average outbreak losses can be in the region of 20 per cent. In addition, ewes which abort can also suffer from a secondary infection of the womb.

During the outbreak the infection can be picked up by mouth by in-contact ewes from the discharge on the ground.

It is considered that the infection can be introduced into clean flocks by wild birds, particularly crows. The ram is not an important transmitter and vibrionic abortion in cattle is not transmitted to sheep. Infected ewes seldom abort again but ewes unaffected one year may abort the next.

Treatment is of little value. All infected foetuses and placentas should be buried or burnt and contaminated ground disinfected as far as practicable. Isolate aborted animals until they have cleaned up. No vaccine is available.

Enzootic abortion (kebbing) is common in southern Scotland and it also occurs to a lesser extent in most parts of England and Wales. It is caused by a large virus which infects both the foetus and placenta resulting in abortion after mid-pregnancy and also the birth of stillborn or weakly lambs at full-term.

There is a diagnostic blood test and this is particularly useful if no other material is available for examination.

Abortion produces some immunity and the disease tends to occur mainly in ewes that have been freshly introduced into the flock. The abortion rate generally runs at approximately 10–12 per cent but in severe outbreaks it can reach up to 40 per cent. Spread occurs at the time of abortion and infection is picked up by the mouth.

Treatment is of little value and prevention lies in efforts to prevent spread by isolation of aborting ewes and the destruction of infected material.

Normal lambs can be born with the infection or, if born uninfected, can pick it up early in life, leading to abortion at their first pregnancy.

Vaccination of uninfected ewes is carried out before service but if the ewe is already infected protection may be delayed until the second pregnancy subsequent to vaccination.

Salmonella abortus ovis, a bacterium which infects the placenta and foetus, is a common cause of abortion in the south-west of England but is tending to spread to other parts of the country. Up to 40 per cent of ewes may abort late in pregnancy but usually the losses are smaller. Death occurs in some ewes following abortion.

Prevention depends on acquiring uninfected ewes and keeping them as a closed flock. There is no effective treatment and no vaccine is available.

Toxoplasma abortion is caused by a small parasite which is often found in a wide variety of animal species including man. Normally, the parasite causes little or no harm but if infection occurs in the developing foetus or in some cases in the very young it can interfere with the normal growth processes, resulting in abortion of the foetus or abnormalities of some organs in the young.

Ewes generally abort during the latter half of pregnancy. The abortion rate is usually not high and immunity following abortion is good. The disease is widespread throughout the country.

There is no treatment and although preventive measures have not yet been adequately worked out, it is suggested that the mixing of uninfected sheep with infected sheep before tupping will give the former immunity and protection before pregnancy.

Border disease is caused by an infectious agent thought to be a virus which can cause serious effect upon the developing lamb in the womb. As well as infertility in both ewes and rams, the disease also causes resorption of the foetus, abortion, and birth of premature lambs. If the lamb survives to term it may be under-weight, weakly, and have a hairy fleece unlike that of its breed. The general conformation of the body is altered, the legs being shorter and the head domed, while a few may show nervous symptoms in the form of continual shaking. The disease can be recognised by these features. Some immunity is developed by the ewe as her subsequent lambs are unusually unaffected.

Other causes of abortion are almost any infection, toxic agent, or stress which upsets the ewe, especially a fever, may result in abortion.

Seventeen Parasites Affecting Sheep

Considerable losses occur in livestock through heavy infestation of parasites. A parasite is a living organism that depends upon a living host for its livelihood. In the case of sheep (host) there are two types of parasite that trouble them:

1 Internal parasites, e.g. roundworms that may live in the lungs, stomach or intestines; flat worms that may live in the liver or intestines.

2 External parasites, e.g. keds, lice, tick, blow-fly, nasal fly, that live on the sheep's skin and in the wool.

Roundworms (Nematodes)

There are about ten different strains of roundworm that are present in the sheep's stomach and intestines, and, providing they are kept in small numbers, will do little damage to the host. It is only when the worm population rises to abnormal levels that suffering takes place. The most troublesome worms are:

Haemonchus contortas—twisted wireworm or barber's pole
Trichostrongylus axei and *vitrinus*, which causes a black scour
Nematodirus filicollis and *battus*, which affects lambs

Roundworms vary in size from tiny thread-like structures about 6 mm long (lungworms) to 300 mm long. They have separate sexes, male and female.

Life cycle

The worms mate, and the female lays fertilised eggs which pass out of the sheep in the droppings. The egg changes into an immature worm, called a larva. The larvae are extremely delicate on hatching and are susceptible to unfavourable weather. They like warm moist conditions, such as we find in the spring. The larvae become infective after 4–7 days from the egg stage, and if picked up by a sheep will develop into a mature roundworm in about twenty-one days from the time the egg was laid.

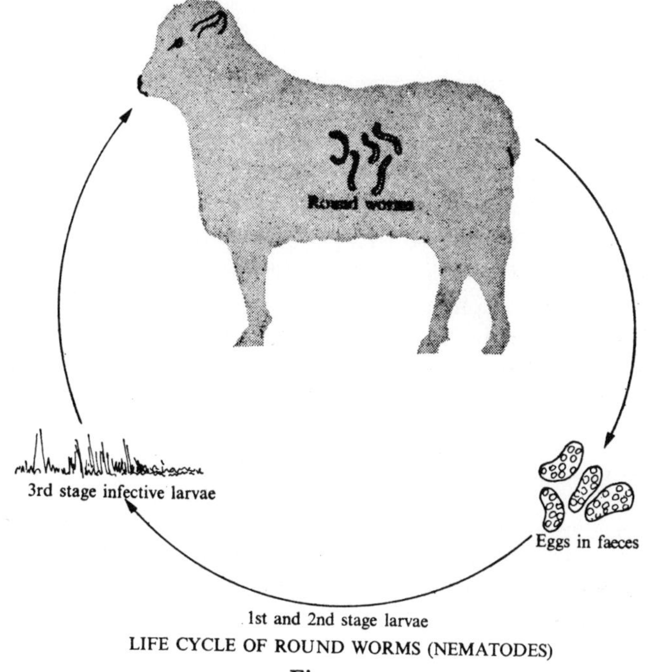

LIFE CYCLE OF ROUND WORMS (NEMATODES)

Fig. 79

Control

If we move the flock on to fresh pastures every 4–7 days we shall reduce the risk of the sheep picking up infective larvae. If it is possible to rest the land from sheep for twelve months, then practically all the larvae will perish. Land ploughed and rested for a year may be considered as perfectly clean.

Other ways of controlling roundworms is to chain harrow the pastures after grazing and to 'top' the long grass with a mowing machine. This will expose the larvae to direct sunlight, which will kill the majority of them.

Mixed grazing with cattle will mean lowering the sheep density per hectare, which in turn will lower the larvae infestation.

Parasitic gastro-enteritis, or simply the infestation of parasitic worms in the stomach, may affect sheep of all ages: but it is most troublesome with young lambs from about six weeks old onwards. The disease frequently affects lambs just after weaning, particularly in July–September when the lamb's diet suddenly changes from milk to all grass. Adult sheep are usually fairly resistant to roundworms, but they should be dosed for worms occasionally to prevent them acting as carriers.

Symptoms

Haemonchus contortas, commonly known as the barber's pole or twisted wireworm, attacks lambs mainly in the spring and throughout the grazing season. The lambs become unthrifty, with a progressive loss of condition; the fleece loses its lustre, and may become broken. On close examination, the membranes of the eye will be found to be pale and anaemic. There is little or no scouring.

Trichostrongylus axei is often called the black scours worm since it causes profuse diarrhoea in infected sheep. This worm is most common during the winter months, especially among root-fattening hoggets. The lambs become stunted and pot-bellied; there may also be a soft swelling under the jaw, which is called 'watery poke'.

Prevention and treatment

Where it is not possible to rotate the grazing, the entire flock should be dosed with an approved anthelmintic.

Dosing programme

Recent research has shown that roundworms reproduce most rapidly during the late spring and decline in the autumn. July–August are the worst months. Routine drenching should be carried out, therefore, during the late spring and through the grazing season.

Dose all ewes 4–5 weeks after lambing.

Dose all lambs when six weeks old, and monthly during April, May, June and July.

Dose ewes in the autumn, about one month before tupping.

N.B. Experimental work has shown that if ewes are dosed 4–5 weeks after lambing, and run on clean pastures, then the lambs will not require dosing.

Nematodirus filicollis

Nematodirus filicollis and *battus* have become a major problem in recent years, especially where sheep are kept intensively. The worm is quite distinct from the other species in that when the eggs are passed to the ground, they can remain dormant for at least twelve months. They hatch out only under favourable weather conditions. Where pastures are grazed continuously there will be a tremendous 'build up' of larvae over one or two years.

Young lambs, 4–6 weeks old, are particularly suceptible to Nematodirus infection, and losses can be extremely high.

Symptoms

Lambs appear dull and lifeless for a few days, and this is followed by profuse scouring. The lambs become weak and death follows very quickly. In acute cases the illness may only last for 2–4 days.

Treatment

Veterinary advice should be sought at the first signs of the disease.

Prevention

There are several excellent anthelmintics now on the market which will give control of the disease.

Where possible, graze the lambs on pastures that have not carried sheep for the previous 12–18 months. One-year leys are ideal in this respect.

Platyhelminthes

The phylum Platyhelminthes consists of two main families that affect man and animals, namely the Cestoda (tapeworms) and the Trematodes (flukes).

The liver fluke (Fasciola hepatica) is one of the most widely distributed and harmful parasites that affect sheep. It is usually associated with wet summers, and in bad years can account for considerable losses in flocks. The earliest symptoms are loss of condition, the gums and eyelids become pale, and a soft watery swelling may be seen under the jaw. A pot-bellied appearance follows, due to an accumulation of fluid in the abdomen. The sheep get weaker, and losses can be very high.

A post-mortem examination shows the liver to be of a hard consistency. If the liver is cut open, the flukes can easily be seen. The liver fluke is about 12 mm–25 mm long and about 12 mm wide, body being quite flat.

Life cycle

In order to complete its life cycle, the liver fluke requires an alternate host to the sheep. In this case the mud snail (limnea truncatula) is the other host.

The mud snail is sharp-pointed, and rarely exceeds 5 mm in length. It is found in permanent and semi-permanent wet-grass areas, and does not usually frequent swift-flowing streams, although the back waters often harbour small colonies of snails.

The essential phases of the life history

1 The adult fluke lays eggs in the bile duct, which pass through the intestines and are removed from the sheep in the droppings. If the eggs fall on dry land, they quickly die, but if dropped in wet areas they can remain alive for 5–6 months, or even more.

2 Between nine days and eight weeks after leaving the sheep the egg hatches into a larva called a miracidium. This swims about vigorously, and within a few hours is picked by by the snail.

3 The miracidium undergoes development inside the snail, and after 6–7 weeks appear in new forms called cercariae. It is important to note that inside the mud snail a second reproduction takes place, the miracidium producing about 1000 cercaria. The cercaria move from the snail and attach themselves to the herbage, where they may remain alive for eight months or more.

4 The grazing sheep eats the grass with the encysted cercaria, and later migrate to the liver, via the blood-stream, and hatch into fully developed liver flukes. This again takes 5–6 weeks. After a further period of 5–6 weeks the flukes mature and begin laying eggs. The process inside the sheep is therefore approximately twelve weeks. Each adult fluke is capable of laying a very large number of eggs, and each egg is capable of developing into 1000 or more carcaria when in the snail. Therefore, one can easily visualise how serious this parasite can be if not strictly controlled.

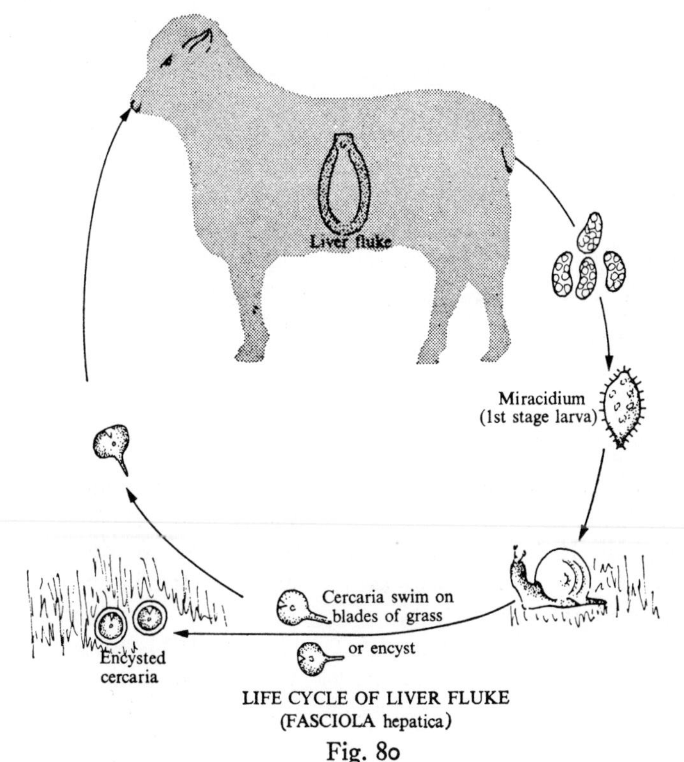

LIFE CYCLE OF LIVER FLUKE
(FASCIOLA hepatica)

Fig. 80

The mud snail (Limnea truncatula)

The mud snail hibernates during the winter months, but becomes active around March–April. The eggs that they produce gives rise to another generation which in turn reproduces a second generation by about July. A third generation is hatched in October. It has been found that a single mud snail in March will be the grandparent of 160000 snails by October.

Prevention and control

1 The same fluke is harboured by cattle, sheep, deer and rabbits. Control measures must therefore be directed at all these forms of livestock when grazing together.
2 By killing the mud snails we can break the life cycle by removing the alternate host.

To destroy the snail:
(i) Spray all water meadows and wet area in June or the autumn with 2 kg–3 kg copper sulphate in 220 litres water per hectare, or mix 2 kg–3 kg copper sulphate in 20 kg sand and broadcast by hand.
(ii) Graze the fields with ducks for they will consume the snails!

Treatment

Drench cattle and sheep with proprietry anthelmintic, or use an approved injection.

Tapeworm (Cestoda)

The tapeworm is a segmented flat worm. It lives as a parasite in all types of livestock, but certain strains affect definite species of animals, e.g. moniezia bendeni affects sheep.

The tapeworm is a very long worm. It starts in the form of a pin head and elongates by segments. These segments are hermaphrodite and therefore self-fertilising. The tapeworm can grow to 300 mm–380 mm long before the segments drop off and pass out of the animal. Two species of tapeworm infect sheep in this country, namely moniezia expensa and moniezia benedeni. They occur in the small intestine of lambs up to about six months of age. They do relatively little damage, unless in large quantities.

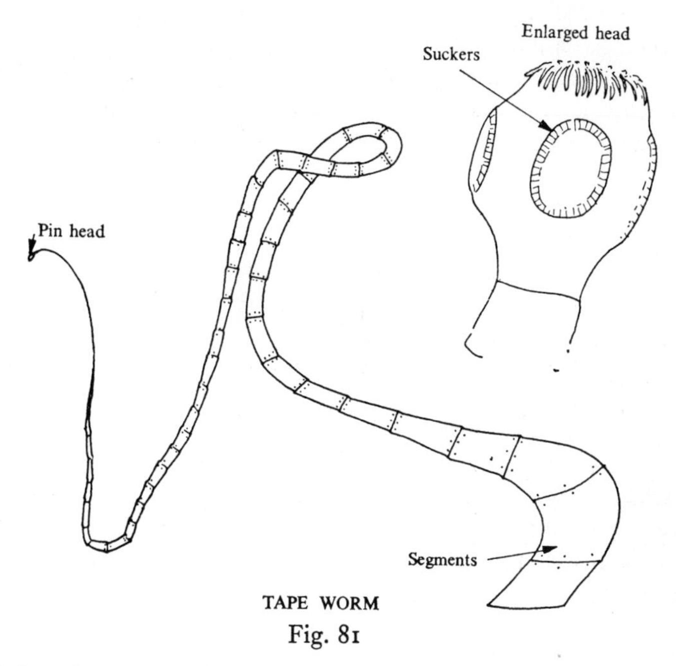

TAPE WORM

Fig. 81

Life cycle

In order to complete their life cycle, tapeworms require an intermediate host, which is a small mite that lives in the grass. The ripe tapeworm segments, which are packed with worm eggs, break off from the ends of the worm whilst in the sheep's bowel and pass out on to the ground in the droppings. The segments disintegrate on the ground and release the eggs, which are then picked up by the mites.

In the mite's body, the eggs develop into an intermediate stage called a cysticeroid or cyst. The lambs, while grazing, pick up the mites containing cysts, and are thus infected with the tapeworm. The tapeworm grows very rapidly inside the lamb, and mature segments will be released in about six weeks after infection.

The most common time for tapeworm is in June to September, but the worms do not last much more than three months, and most of them disappear in the late autumn.

Symptoms

Only heavy infestations of tapeworms do noticeable harm to the lamb. Occasionally a death occurs due to two or more worms blocking the bowel and causing a blockage.

Heavy infestations cause 'pot-bellied' lambs, and general unthriftiness.

Lambs over six months of age do not usually suffer from tapeworms.

Treatment

Drench the lamb with proprietry anthelmintic.

Tick, keds, lice and mites

Ticks, keds, lice and mites are all parasites that live in the wool of the sheep. They all cause severe irritation and discomfort to the sheep; some suck blood, whilst others live on secretions from the skin. Ticks are most troublesome, since they are able to carry two serious diseases: tick borne fever and louping-ill.

Fortunately all these parasites may readily be controlled by either spraying or dipping the entire flock in an approved sheep dip.

Eighteen Sheep Records and Accounts

In order that a flock may be both profitable and efficient, it is essential that proper breeding records, and some form of cost accounts, are kept. This will enable the farmer to judge the performance of his sheep and provide valuable information when making future budgets or changing the farm policy.

Breeding records

The first essential is to be able to positively identify each individual sheep in the flock. This can be achieved by either ear-marking using a tattoo, ear tag, or ear notch or by using paint brand numbers on the sheep's side. In the case of horned breeds, horn branding may be adopted.

Pedigree flocks must use the ear-tattoo system, which is the only really permanent mark; ear tags sometimes pull out and notched ears occasionally get torn. Many commercial breeders use a combination of ear tags and paint branding—the large numbers on the side are clearly distinguishable and especially useful at lambing time when a ewe may be identified without having to catch her to read the tattoo or ear tag (see Figs. 17, 18, 19 and 20, page 41 and 42).

Pedigree flock book—breeding register

Each Flock Book Society provides its members with their own specially designed pro-forma, on which the farmer records such information as the pedigree of sheep, date of birth, ear marks, sire/dam and in the case of breeding ewes the sire of offspring, number and sex of lambs, date of lambing, etc. A remarks column is included for such items of disposal of lambs and prizes won at shows.

PEDIGREE ...... EWE BREEDING RECORD

EAR MARK NUMBER E 121	BREEDER John Smith	SIRE Hope King	DAM B.96
FLOCK BOOK NUMBER 57641	Hope Farm	E.M. D.172	E.M. B.96
		Flock Book 35422	Flock Book 23769

Wool Weight		1967-2·3kg	1968-2·7kg	1969-2·7kg	1970-2·5kg	19 kg	19 kg	19 kg	19 kg
Year	Ram E.M.	Ewe E.M.	Date of Birth	Weight at Birth	Weight at 1 Month	Weight at Weaning	Weight at Sale	Sire F.B.	Date Sold £ p.
1968	G·22	G 23	10th March	4·5 kg 4·1 kg	13·6 kg 12·7 kg	36·1 kg 36·1 kg	46·1 kg Sold fat	Hope Duplex 57682	10/10/ 25·00 30/6/ 7·50
1969	H·19		15th Feb	5·9 kg	15·9 kg	39 kg	46·7 kg	Hope Duplex 57682	11/10* 30·00
1970	I 30	I 31	20th Feb	5·0 kg 4·5 kg	13·6 kg 13·6 kg	36·3 kg 36·3 kg	Sold fat	Hope Dan 62631	15/6 8·20 8·20
19									
19									
19									
19									

REMARKS: * 1st prize Ram lamb @ show and sale

Fig. 8[illegible]

LAMBING RECORD SHEET

EWES DUE TO LAMB 9TH – 23RD MARCH 1970 –

BLUE MARK

Steaming up Feb 1st 1970. Vaccine March 3rd 1970

Breed	E.M.	Sire	Date Lambed	M E.M.	F E.M.	Kg			Remarks
Kerry	1	Robin (R)	9/3	1	2	4.5 / 4.1			
"	5	Robin	10/3	3	—	6.4			
"	8	Craftsman (C)	10/3	4	—	6.4			
"	13	Craftsman	14/3	9	10	4.5 / 4.5			Drew lamb – Vet:
"	16	Craftsman	17/3	13	—	5.9			
"	30	Model Boy (M)	11/3	5	6	4.1 / 5.0			
Masham	35	(M)	21/3	14	15	4.5 / 4.5			
"	37	(M)	12/3	11	12	5.0 / 4.1			
"	38	(M)	11/3	7	8	4.5 / 3.6			
	39	(M)							

Fig. 83

Commercial records

The commercial breeder does not require such comprehensive records as the pedigree breeder. A simple record sheet is shown on page 201, the essential features being date of tupping, date when lambs are due, date for steaming up. The individual ewe's record will show number of lambs born, sex and ear number of lamb, and perhaps the weight of lamb at birth. Columns are included for subsequent weighing of lambs, and the date of disposal.

COMMERCIAL FLOCK RECORD SHEET

Flock 'Made up' on ......14th Sept...... 1977

No

Breeding ewes – 1 years ..25 Cluns..........
2 years ..25 Cluns..........
3 years ..20 Masham....
4 years ..30 Masham....

Stock ewes ......Nil..................

Breeding Rams 1 Suffolk
..1 Ryeland.........
....
Store lambs ..Nil.......

Total ewes put to ram 100.............................

Ram turned with flock ..1st Oct....1977

No marked		due to lamb
yellow	73	Feb 27th 1978
red	30 (3 returns)	March 15th 1978
blue	4 (all returns)	April 1st 1978

Ram removed from flock 21st Nov 77   Lambing ends April 19th 1978

Lambing percentage: $\dfrac{\text{Number of lambs reared } 171}{\text{Number of ewes put to ram } 100} \times \dfrac{100}{1} = 171\%$

Lambing percentage last year ..168....%

Sale of Sheep

No		£ . p
Ewes and lambs		
Cull ewes		
Fat lambs	150 @ £25	3,750
Store lambs	21 @ £20	420
Other sheep		
Total sales		£4,170

Total weight of wool ..261.. Kg
Average weight of fleece 2·68.. Kg
Sale Price Total ....£250......p.
Per fleece £2·50.....

Fig. 84

Movement of Animals (Records) Order, 1960

Date of Movement	Particulars of each bovine animal moved to or moved from premises mentioned on front cover				Number of sheep, goats or pigs (specifying which)	Movements to premises mentioned on front cover	Movements from premises mentioned on front cover
	Breed	Age	Sex	Ear Mark or Ear Tag. No.		Premises from which moved (including Market, Saleyard or Fair) *and/or* Name and Address of person from whom delivery was taken	Premises to which moved (including Market, Saleyard or Fair) *and/or* Name and Address of person taking delivery.
						N.B. Both of these particulars are to be entered if available	N.B.– Both of these particulars are to be entered if available

Fig. 85

It can be readily appreciated that ewes that produce very small lambs or are poor milkers can be quickly detected by glancing through the record sheet. Barren ewes and those with bad udders may be noted on the sheets, and disposed of as soon as possible.

Livestock movement register

All livestock owners must, by law, keep a register of the movement of cattle, sheep and pigs. This is necessary in order that the police and veterinary surgeons can trace livestock after they have left the farm should there be an outbreak of one of the notifiable diseases.

The police have powers to call on a farmer and ask to see his movement book, at any reasonable time.

Enterprise costs and budgeting

Sheep are, perhaps, the most difficult farming enterprise to cost accurately. On many farms they are kept as scavengers, whilst on others they make considerable demands on the best feeding pastures. It must be emphasised that the following examples, although taken from actual sheep costings, are intended as a guide only. The farmer must apply his own costs, which are applicable to his system, and operate under the local conditions. Nevertheless the reader should gain some advantage from a discussion on the various factors that influence the costs, outputs and profitability.

Factors affecting the cost of maintaining a breeding flock

Grazing — Supplementary food
Labour — Depreciation of ewe
Stocking rate — Veterinary and medicine
Miscellaneous, including service—shearing—haulage.

Grazing

Grazing costs vary considerably from farm to farm. This is due mainly to the rent, amount of fertilisers used, whether the grazing is permanent or temporary pasture, the natural fertility of the soil and the stocking rate.

It should not be assumed, however, that because a certain farm has a low rent the grazing costs are automatically low. You may pay a higher rent for good quality pastures, but if they support eighteen ewes and their lambs per hectare during the grazing season the individual cost per sheep may be less than on cheaper land, with correspondingly lower stocking rate.

There is considerable difference of opinion amongst farmers as to how much fertiliser should be used on sheep pastures. Some rely entirely upon the application of 500 kg per hectare of basic slag every three years, whilst others have been known to use up to seven hundred units of nitrogen per hectare per annum.

A fairly safe generalisaion is to suggest that sufficient fertiliser should be applied to just keep the grass growing ahead of the flock. Very heavy dressings of nitrogenous fertilisers can lead to wastage of grass through soiling, i.e. (heavy stocking will mean many feet trampling the grass) and may also cause the ewes to scour.

The main items of expense in grazing are rent, seeds, manures, maintenance of fences, ditches and hedges. These on the lowland farm at the time of writing amount to around £40–£50 per hectare. If we stock 12 ewes per hectare, the cost per ewe will be about £3·50–£4 per ewe per annum.

Stocking rate

In order to work out the number of ewes kept per hectare throughout the year, it is necessary to calculate the amount of winter feed that each ewe will require in addition to the summer grazing.

Taking a hypothetical case one can argue that if a ewe was fed indoors from say December (end of grazing season) until the end of March and fed 1·3 kg hay or its equivalent per day for the four-month period, she will require at least 150 kg of hay, which must be produced off the non grazing land.

One could assume that a hectare of grass would produce the equivalent of 7·5 tonne of hay throughout the growing season. Therefore, one hectare of land will support:

$$\frac{\text{7·5 tonne hay or hay equivalent (e.g. silage)}}{\text{150 kg hay or H.E. per ewe}} = \text{50 ewes or 0·02 ha per ewe}$$

If we stock our land at 18 ewes per hectare for summer grazing, each ewe will require 0·055 ha, making a total of 0·057 ha per annum. Therefore, 1 hectare will support 13 ewes per hectare throughout the year.

Winter feed—requirements per ewe		*Summer grazing (Ewes per hectare)*		*Annual stocking rate per hectare*
0·02 ha	+	10	=	8·3 ewes ha
0·02 ha	+	12·5	=	10·0 ewes ha
0·02 ha	+	15·0	=	11·5 ewes ha
0·02 ha	+	17·5	=	12·5 ewes ha
0·02 ha	+	20·0	=	15·0 ewes ha

Winter grazing

In practice we find that many flocks, especially those descended from the hill breeds, will find a fair amount of food, if allowed to roam over stubbles and clean up pastures. If possible allow about 0·4 ha per ewe for the winter period, and offer supplementary hay once the grass becomes scarce.

Supplementary food—hay and concentrates

Obviously it is difficult to state exactly how much hay a ewe will consume during the winter months, but as a guide we may reckon on feeding 0·45 kg per head from December to March, providing a reasonable amount of forage is available. The concentrate allowance will depend mainly on the time of lambing. Early lambs must be fed generously in order that they grow quickly to catch the Easter market, whilst April-born lambs will receive little other than milk and grass.

e.g. Ewe—fed 0·45 kg hay per day
= 12 kg per month
If fed 4 months = 50 kg

Concentrates:
Steaming up 12·5 kg
Lactation 0·45 kg/day 6 weeks = 19 kg
Lamb(s) 0·45 kg/day 4–6 weeks = 12 kg–19 kg
Approximately 36 kg–50 kg per ewe
Cost: Hay 50 kg at £40 tonne = £2
Concentrate 36 kg–50 kg at £100 tonne = £3·60–£5

An approximate figure would be £6 per ewe.

Ewe depreciation

Ewe depreciation is the difference in price between buying young ewes and selling cull ewes, plus any mortality.

e.g. Purchase 100 yearling ewes at £50 each = £5000
Sale 95 five-year-old ewes at £25 each = £2375
after four crops of lambs and allowing 5 per cent mortality.

$$\text{The ewe depreciation} = \frac{£2625}{100 \text{ ewes}} \div 4 \text{ years}$$

$$= £26{\cdot}25 \text{ per ewe or } £6{\cdot}37 \text{ per ewe per annum}$$

The depreciation price will vary according to the longevity and liveability of the flock, and the buying and selling price.

Labour cost

This is quite an easy cost to assess where a full-time shepherd is employed to look after a large flock, e.g.

$$\frac{\text{Shepherd's wages} + \text{house} = £3000}{600 \text{ ewes}} = £5 \text{ per ewe}$$

In small flocks it is more difficult to cost. Many farmers keep 60–80 ewes and look after them themselves. In such a case it would be very difficult to put an actual cost, but for budgeting purposes we should think of a similar cost per ewe as in larger flocks—say £4–£5 per ewe.

Ram—service cost

This expense is calculated by dividing the annual depreciation of the ram by the number of ewes mated to him.

Purchase of ram at say £100
Sale of ram after four seasons = £25

Depreciation = £75

$$\text{Annual depreciation} = \frac{£75}{5 \text{ years}} = £25$$

Cost of service if mated to sixty ewes = 40p per ewe. The cost of keeping a ram throughout the year would be offset, more or less, by the sale of his wool.

Veterinary and medicine

Veterinary expenses are usually fairly low with sheep. There may be the occasional difficult lambing and from time to time it may be necessary for a Caesarian operation to be performed, but in general the vet's charges are unlikely to be more than a few pence per head.

Medicines, in the form of worming doses, vaccination against clostridial diseases, and foot-rot treatment can amount to a further 25–30p per head.

A fair figure for vet and medicine would, therefore, be about 50p per ewe per annum.

Shearing

The present-day contractor's charge for shearing is between 28p and 30p per head. The farmer is expected to find a man to catch the sheep and roll the fleece. The farmer may, of course, shear his sheep, but there will be a cost for machinery and sharpening combs and cutters.

Overheads

Farm 'overheads' is a term that refers to all the traditional costs on a farm such as telephone, electricity, farm transport, repair and maintenance of roads, gates and ditches. These costs are unlikely to exceed £6 a hectare, and could be much less. If we stock twelve ewes per hectare this means 50p per head.

Outputs or sales

The main income from the flock will be the sale of lambs and wool, occasionally there will be cull ewes and rams. Lambs marketed around Easter will command a high price per kilogram, as the season progresses and the summer 'grass-fed lambs' come on to the market the prices will fall. This means that it may be most profitable to sell young lightweight lambs at around 31 kg l.w. at the higher price early in the season, rather than keeping the lamb until it reaches 36 kg when the price may have fallen. One cannot be dogmatic on this point because prices fluctuate from season to season and according to the supply and demand.

Example—costs and returns

Costs	£ p	*Returns*	£ p
Grazing	3·50	Lambs:	
Supplementary food:		150% at £25	37·50
Hay	2·00	Wool 3 kg at £1 per kg	3·00
Concentrates	3·50		
Ewe depreciation	6·00		
Labour	5·00		
Ram—service	·40		
Shearing	·30		
Vet and medicine	·50		
Overheads	·50		
Total costs	21·70		
Profit	18·80	Loss	
	£40·50		£40·50

If ewes are stocked at 12 per hectare a profit of over £120 per hectare would be possible.

APPROXIMATE CONVERSION TABLES–READY RECKONER

WEIGHT

kilo-grammes (kg)		*pounds (lb)*	*pounds (lb)*		*kilo-grammes (kg)*	*kilo-grammes (kg)*		*hundred-weights (cwt)*	*hundred-weights (cwt)*		*kilo-grammes (kg)*
0·5	=	1·10	0·5	=	0·23	25	=	0·49	0·5	=	25·4
1	=	2·20	1	=	0·45	50	=	0·98	1	=	51
2	=	4·41	2	=	0·90	100	=	1·97	2	=	101
3	=	6·61	3	=	1·36	150	=	2.95	3	=	152
4	=	8·82	4	=	1·80	200	=	3·94	4	=	202
5	=	11·00	5	=	2·26	250	=	4·92	5	=	254
6	=	13·20	6	=	2·70	300	=	5.90	6	=	305
7	=	15·40	7	=	3·10	350	=	6·89	7	=	355
8	=	17·60	8	=	3·60	400	=	7·87	8	=	404
9	=	19·80	9	=	4·00	450	=	8·86	9	=	456
10	=	22·00	10	=	4·50	500	=	9·84	10	=	508

AREA

hectares (ha)		*acres (acres)*	*acres (acres)*		*hectares (ha)*
0·5	=	1·24	0·5	=	0·2
1	=	2·47	1	=	0·4
2	=	4·94	2	=	0·8
3	=	7·41	3	=	1·2
4	=	9·88	4	=	1·6
5	=	12·36	5	=	2·0
10	=	24·71	10	=	4·0

square metres (m^2)		*square feet (ft^2)*	*square feet (ft^2)*		*square metres (m^2)*
1	=	10·76	1	=	0·09
2	=	21·53	2	=	0·18
3	=	32·29	3	=	0·27
4	=	43·06	4	=	0·36
5	=	53·82	5	=	0·46
10	=	107·62	10	=	0·92

LENGTH

millimetres (mm)		*inches (in)*	*inches (in)*		*millimetres (mm)*
25	=	0·99	1	=	25·4
50	=	1·97	2	=	50·8
100	=	3·94	3	=	76·2
200	=	7·87	4	=	102·0
300	=	11·80	5	=	127·0
400	=	15·70	6	=	152·0
500	=	19·70	12	=	305·0

metres (m)		*yards (yd)*	*yards (yd)*		*metres (m)*
0·5	=	0·55	0·5	=	0·46
1	=	1·09	1	=	0·91
2	=	2·19	2	=	1·8
3	=	3·28	3	=	2·7
4	=	4·37	4	=	3·6
5	=	5·47	5	=	4·5
10	=	10·90	10	=	9·1

CAPACITY

litres (litre)		*gallons (gall)*	*gallons (gall)*		*litres (litre)*
0·5	=	0·11	0·5	=	2·27
1	=	0·22	1	=	4·55
2	=	0·44	2	=	9·09
3	=	0·66	3	=	13·64
4	=	0·88	4	=	18·18
5	=	1·10	5	=	22·73
10	=	2·20	10	=	45·46

Glossary of Terms

Sheep

Barren ewe	Empty ewe that failed to breed
Cull ewe	A ewe removed from the breeding flock because of some defect, e.g. faulty udder
Draft ewe	Breeding ewe, still sound in feet, teeth and udder, but may not be suitable for keeping in the flock This term is associated with hill flocks, where draft ewes are sold to lowland farms for further breeding
Ewe	Female sheep kept for breeding
Ewe lamb or theve lamb	Female lambs intended for breeding, either as lambs or later as yearling ewes
Fat lamb—butcher's lamb	Male or female lamb intended for slaughter
Full mouth, four-year-old ›	Fourth to fifth shearing, all have eight broad teeth
Hoggets, hog	Male or female over six months of age intended for slaughter
Ram, tup, tip	Male sheep usually pure bred to pedigree kept for breeding—between six to twelve months
Ram lamb or tup lamb	Male sheep intended for breeding
Shearing, shearling, yearling, two tooth, shear hog, wether gimmer	First to second shearing. Most commonly called yearlings, all have two broad teeth showing
Stock ewes	Ewes over five years old that are still kept on for breeding, probably because they are particularly valuable
Tegs, wether lamb, wedder lamb, wedder hog, ram hog, tup teg	Castrated male sheep intended for slaughter. 12 months old

Three-year-old 3 shear	Third to fourth shearing, all have six broad teeth
Two-year-old 2 shear	Second to third shearing, all have four broad teeth showing

Shepherds' and Wool Terms

Aged	Sheep past their best—may have lost some teeth—usually six-year-olds and over
Anaerobic	The ability to live without free oxygen, e.g. Clostridium tetani which causes tetanus in humans and farm animals is an anaerobic bacteria—can live in the soil without free oxygen
Ash	Mineral matter or minerals found in feedingstuffs
Barley lamb	Lambs housed and fed with an all concentrated diet for fattening
Bradford spinning count	This indicates the number of hanks of wool that can be spun from 450 g of wool. A hank is 510 metres. The higher the number the finer the wool e.g. 60s indicates fine hosiery wool 36s indicate coarse wool for carpets
Branding	Paint marks used for identifying sheep, e.g. branding the farmer's initials on the sheep's side
Break wool	A temporary interference in wool growth causing a marked thinning in the wool fibre—breaks occur if sheep are kept in a semi-starved state for short periods, or after a difficult lambing
Breech presentation	Lamb born with the hind legs and tail presented first
Brightness wool	A term used to describe a light reflecting quantities of white wools
Canary stain	Bright yellow stain in fleece—detracts the value of wool—cannot be removed by normal washing

Card and carding	A process of teasing out clumps of wool fibres to run parallel to each other
Castration	Removal of the testicles by surgery, or by crushing the spermatic cord with a burdizzo instrument
Condition	Indicates the degree of finish or fatness in sheep
Counts—wool	Indicates wool quality—see Bradford spinning count
Couples	Ewe and lamb
Creep	Pen or field where lambs may 'creep' to gain extra food, and ewes cannot gain access
Crimp wool	The natural wave in wool
Cross-bred	The offspring resulting from mating two or more breeds
Cuckoo lambs	Lambs born after mid-April
Cull—culling	Sheep considered unsuitable for breeding, e.g. culled because of damaged udder
Dagging—burling—belting	Removal of dirty wool from around tail area
Dam	Female maternal parent—used in sheep's pedigree
Dandy brush	Strong, stiff bristled brush used to remove dirt from fleece when preparing sheep for 'show'
Dead wool	Wool clipped from dead sheep before carcass is disposed of
Density wool	Denotes compactness of wool growing on the sheep, e.g. the close proximity of fibres grown on a given surface area
Depth of staple	Length of wool, measured from sheep's skin to outer tip of fleece
Development	Change in body shape or outline
Doubles	Ewe with twin lambs
Dry ewe	Ewes that are not in milk

Ear mark	A distinctive mark, either tattoo, or ear notch—used to identify sheep
Ear tag	A metal, plastic or nylon tag used to positively identify individual sheep
Ear tattoo	Number tattooed into ear using needles and tattoo paste
Elasticity wool	The ability of wool fibres to return to their original form after being stretched or compressed
Embryo	The unborn lamb
E.D.C.W.	Estimated dead carcass weight. When lambs are sold for slaughter they are weighed alive, and the grader by his experience and handling of sheep is able to estimate the dead carcass weight which is usually about 48–51% of the live weight
Fellmongering	Removal of wool from the skin after sheep are slaughtered
Fleece	The sheep's coat of wool
Flock	A number of sheep
Flushing	Feeding breeding ewes extra before tupping in order to encourage the ovaries to produce more eggs
Fly strike	Maggots in the skin and wool
Fly-blown	Caused by the blow-fly laying eggs on the sheep
Gestation	Pregnancy—from conception to birth—usually referred to as in-lamb
Gestation period	147–151 days
Gummy	An aged sheep that has lost all its incisor teeth
Half–bred	The progeny of two distinct breeds, e.g. Border Leicester × Cheviot produces Scots Half–bred
Hank-wool	510 m of wool—see Bradford spinning count

In-breeding	Mating sheep closely related to one another, e.g. father to daughter
In-bye	The fenced area of a hill sheep farm
In-lamb	A ewe that is pregnant
Kemp	A hard brittle fibre similar to hair —usually associated with hardy hill breeds, and frequently found in the breech area of the fleece
Killing out percentage—KO%	The yield of carcass, e.g. carcass weight 15 kg, live weight 30 kg $\frac{15}{30} \times \frac{100}{1} = 50\%$ killing out percentage
Lactation	The period a ewe suckles her lamb
Lamb	A young sheep under six months
Lamb—meat	The flesh of young sheep
Lambing percentage	The number of lambs *reared* per 100 ewes put to ram
Lambing potential	The number of lambs *born* per 100 ewes put to the ram
Line breeding	Breeding from sheep of the same family, but not close relations
Lot	A parcel of wool or a group of sheep offered for sale by auction—hence Lot number . . .
Lustre wool	The glossy appearance of wools
Make-up	The deficiency payment made by the Government under the Fat Stock Scheme
Making-up the flock	Selection of ewes and rams for mating
Masham	Teeswater or Wensleydale × Swaledale or Dales bred
Mating	The act of copulation sometimes referred to as service
Maturity	Full development of skeleton muscle and fat
Mule	Blue-faced Leicester × Swaledale

Muscle	Flesh or red meat
Mutton	Edible flesh of mature sheep
Oestrum	The period when a ewe will mate with a ram
Overshot	A condition where the sheep's lower jaw extends further than the upper dental pad
Parrot mouth	A condition where the central incisor teeth are crossed in front of another
Pedigree	Record of ancestry
Pelt	A sheepskin with little or no wool on it
Pinch or pinched	Refers to lambs castrated by the Burdizzo method
Prepotency	The power of an individual animal to transmit its characteristics upon a large proportion of its progeny
Pure–bred	A sheep of a recognised breed bred pure for many previous generations
Quality	Standard of excellence
Rig	Male sheep with one undescended testicle or male sheep not properly castrated
Roomy ewes	Big framed ewes of good conformation
Scour	Diarrhoea
Scour wool	Wool washed with soap and water
Scrotum	The purse or bag containing the ram's testicles
Sebaceous glands	The skin glands that secrete wool wax or 'yolk'
Second cuts	Cutting through the middles of the staple of wool, and then 'second' cutting to shear the wool cleanly
Semen	Fluid produced by the male containing sperms
Sire	Male animal—parental parent

Starch equivalent	Measurement of energy in foodstuffs
Steaming-up	Feeding supplementary concentrates to in-lamb ewes prior to lambing
Sterility	The inability to reproduce offspring
Stilboestral	Hormone found in the female—synthetic stilboestral may be implanted into hoggs to improve growth rate
Store lamb	Lambs in good healthy condition, but not yet ready for slaughter
Synthetic	Artificial production of chemical produce a substance, e.g. synthetic oestrogens—Hexoestral and Stilboestral
Testicles	Sperm-producing glands. Normally the male has two testicles suspended in the scrotum
Tupping time	The time when the ewe is mated with the ram or tup
Yolk	Secretion from the sebaceous glands

Bibliography

BAKEWELL, R., *Observations on the Influences of Soil and Climate Upon Wool*, Harding, London, 1808

BELSCHNER, H. G., *Sheep Management and Diseases*, 8th ed, Angus and Robertson, London, 1965

BOWEN, G., *Wool Away! The Technique and Art of Shearing*, 2nd ed, Whitcombe and Tombs, Auckland, N.Z., 1965

BRITISH VETERINARY ASSOCIATION, *Handbook on Meat Inspection*, British Veterinary Association, London, 1965

BRITISH WOOL MARKETING BOARD, *British Sheep Breeds; Their Wool and its Uses*, British Wool Marketing Board, Isleworth, 19

CLARKE, H. G., *Commercial Sheep Management*, Crosby Lockwood, London, 1963

CLARKE, H. G., *Practical Shepherding*, Farmer and Stockbreeder, London, 1959

COLE, V. G., *Sheep Management for Wool Production*, Grazcos Co-operative, Sydney, 1963

COOPER, M. Mc., and THOMAS, R. J., *Profitable Sheep Farming*, Farming Press (Books) Ipswich, 1965

COWLEY, C. E., *Classing the Clip; A Handbook on Wool-Classing*, 4th ed, Angus and Robertson, Sydney, 1944

DEPARTMENT OF AGRICULTURE AND FISHERIES FOR SCOTLAND, *The Feeding of Farm Animals*, H.M.S.O., Edinburgh, 1966

ENSMINGER, M. E., *Sheep and Wool Science*, Danville, Illinois, Luterstake, 1964

EVANS, R. E., *Rations for Livestock*, Ministry of Agriculture, Fisheries and Food, H.M.S.O., London, 1966

FARMERS' WEEKLY, *The Shepherds Guide*, Farmers' Weekly, London

FRASER, A., *Sheep Farming*, 5th ed, Crosby Lockwood, London, 1950

GREIG, J. R., *The Shepherd's Guide to Prevention and Control of Diseases of the Sheep*, H.M.S.O., Edinburgh, 1953

HAFEZ, E. S. E., *Reproduction in Farm Animals*, 2nd ed, Lea and Febiger, Philadelphia, 1968

HALNAN, E. T., and GARNER, F. H. A. Eden, *The Principles and Practice of Feeding Farm Animals*, 5th ed, Estate Gazette Ltd, London, 1946

HAMMOND, J., *Farm Animals; their Breeding, Growth and Inheritance*, 3rd ed, Edward Arnold, London, 1960

HAMMOND, Sir John, *Animal Breeding*, Edward Arnold, London, 1963

HAMMOND, J., *Progress in the Physiology of Farm Animals*, Butterworths, London, Vol. 1: 1954, Vol. 2: 1955, Vol. 3: 1959

H.M.S.O., *The Structure of Agriculture*, Ministry of Agriculture, Fisheries and Food, Scotland

JACKSON, B. G., *The Economic Position of Sheep in the Eastern Counties: A Report on a Survey Concerning the Years 1961–1964*, University School of Agriculture Farm Economics Branch, C.U.S.A., Cambridge, 1965

JEFFERY, R. R., Breeding flocks in the upland areas of Cotswood and Wiltshire Downs, Report No. 5, *Some Aspects of the Sheep Industry in the West of England*, Department of Economics, Bristol University

JENNINGS, J., *Feeding, Digestion and Assimilation in Farm Animals*, Pergamon, Oxford, 1965

JONES, E., *Just Your Meat or the Judging of Meat Animals*, 2nd ed, Headley Bros., London, 1955

KAMMLADE, W. G., and KAMMLADE, W., *Sheep Science*, Lippincott, Chicago, 1955

LAWRIE, R., *Meat Science*, Pergamon, Oxford, 1966

LINE, E., *The Science of Meat and Biology of Food Animals*, Meat Trade Journal, 2 vols, London, 1932

McCOOPER, *Taking Sheep off the Hills*, Agriculture, vol. 70, no. 10.

McMEEKAN, C. P., *The Principles of Animal Production*, 2nd ed, Whitcombe and Tombs, London, 1934

MARSHALL, F. H., and HALMAN, E. T., *The Physiology of Farm Animals*, 4th ed, Cambridge University Press, 1948

MINISTRY OF AGRICULTURE, *Committee on Sheep Recording and Progeny Testing*, H.M.S.O., London, 1961

MINISTRY OF AGRICULTURE, *Sheep Breeding and Management*, 2nd ed, H.M.S.O., London, 1960

MINISTRY OF AGRICULTURE, *Fatstock Guarantee Scheme 1967/68*, H.M.S.O., London, 1967

NELSON, R. H., *An Introduction to Feeding Farm Livestock*, Pergamon, Oxford, 1964

NEWSOM, I. E., *Sheep Diseases*, 3rd ed, by Hadleigh Marsh, Maryland, William and Wilkins, Baltimore, 1965

ONIONS, W. J., *Wool: An Introduction to its Properties, Varieties, Uses and Production*, Benn, London, 1962

PAWSON, H. C., *Robert Bakewell: Pioneer Livestock Breeder*, Crosby Lockwood, London, 1957

PIERCY, W. PLANT, *A Practical Treatise on the Merino Sheep and Anglo Merino Breeds of Sheep*, London, 1809

, *Wool Marketing*, Pergamon, Oxford, 1966
Diseases of Sheep, Ministry of Agriculture,
lletin No 170, H.M.S.O., London, 1966
rition, Macmillan, London, 1955
Production and Grazing Management,
n, 1965
ow Your Farm Stock, Scottish
ber, London, 1966
University Press,
ll, London,
tter-

Index